Documentation for the State Variables Package for the Groundwater-Management Process of MODFLOW-2005 (GWM-2005)

By David P. Ahlfeld,[1] Paul M. Barlow,[2] and Kristine M. Baker[1]

[1]University of Massachusetts
[2]U.S. Geological Survey

Techniques and Methods 6–A36

U.S. Department of the Interior
U.S. Geological Survey

U.S. Department of the Interior
KEN SALAZAR, Secretary

U.S. Geological Survey
Marcia K. McNutt, Director

U.S. Geological Survey, Reston, Virginia: 2011

For more information on the USGS—the Federal source for science about the Earth, its natural and living resources, natural hazards, and the environment, visit http://www.usgs.gov or call 1-888-ASK-USGS

For an overview of USGS information products, including maps, imagery, and publications, visit http://www.usgs.gov/pubprod

To order this and other USGS information products, visit http://store.usgs.gov

Preface

This report describes a new package for the Groundwater-Management (GWM) Process for the 2005 version of the U.S. Geological Survey modular three-dimensional groundwater model, MODFLOW-2005. The new package—the State Variables Package—allows a user to specify head, streamflow, or change-in-aquifer-storage state variables in a groundwater-management simulation. The performance of the program has been tested in a variety of applications, some of which are documented in this report. Future applications, however, might reveal errors that were not detected in the test simulations. Users are requested to notify the U.S. Geological Survey of any errors found in this report or the computer program by using the address on the inside of the back cover of the report. Updates might occasionally be made to both the report and to the computer program. Users can check for updates on the Internet at http://water.usgs.gov/software/lists/groundwater/.

Contents

Figures

Table

Conversion Factors

Multiply	By	To obtain
cubic foot (ft^3)	0.02832	cubic meter (m^3)
cubic foot per day (ft^3/d)	0.02832	cubic meter per day (m^3/d)
foot (ft)	0.3048	meter (m)
foot per day (ft/d)	0.3048	meter per day (m/d)
square foot per day (ft^2/d)	0.09290	square meter per day (m^2/d)

Abbreviations

Symbol	Explanation
GWF	Groundwater Flow (Process)
GWM	Groundwater Management (Process)
LP	linear programming
LGR	local grid refinement
SLP	sequential linear programming
STA	State Variables (Package)

Documentation for the State Variables Package for the Groundwater-Management Process of MODFLOW-2005 (GWM-2005)

By David P. Ahlfeld, Paul M. Barlow, and Kristine M. Baker

Abstract

Many groundwater-management problems are concerned with the control of one or more variables that reflect the state of a groundwater-flow system or a coupled groundwater/surface-water system. These system state variables include the distribution of heads within an aquifer, streamflow rates within a hydraulically connected stream, and flow rates into or out of aquifer storage. This report documents the new State Variables Package for the Groundwater-Management Process of MODFLOW-2005 (GWM-2005). The new package provides a means to explicitly represent heads, streamflows, and changes in aquifer storage as state variables in a GWM-2005 simulation. The availability of these state variables makes it possible to include system state in the objective function and enhances existing capabilities for constructing constraint sets for a groundwater-management formulation. The new package can be used to address groundwater-management problems such as the determination of withdrawal strategies that meet water-supply demands while simultaneously maximizing heads or streamflows, or minimizing changes in aquifer storage. Four sample problems are provided to demonstrate use of the new package for typical groundwater-management applications.

Introduction

The Groundwater-Management (GWM) Process provides a set of linear, nonlinear, and mixed-binary (integer) optimization-modeling techniques for the MODFLOW groundwater model that can be used to solve several types of groundwater-management problems. These problems include limiting groundwater-level declines, streamflow depletions, and land subsidence; managing groundwater withdrawals; and conjunctively using groundwater and surface-water resources. The initial release of the GWM Process (Ahlfeld and others, 2005) was designed for the 2000 version of MODFLOW (Harbaugh and others, 2000) and is referred to as GWM-2000. GWM-2000 was later modified to be compatible with MODFLOW-2005 (Harbaugh, 2005) and the local grid refinement (LGR) capability of MODFLOW-2005 (Mehl and Hill, 2005 and 2007). This updated version of the software is referred to as GWM-2005 (Ahlfeld and others, 2009).

Both versions of GWM provide for the control of groundwater levels (hydraulic heads), drawdowns, and hydraulic gradients through the use of head constraints, and for the control of streamflow and streamflow depletions through the use of streamflow constraints. Neither version of GWM allows for direct control of aquifer storage, although changes in aquifer storage can be managed indirectly through the use of head constraints. This report documents a new capability for GWM-2005 that extends the current options to allow control of heads, streamflows, and changes in aquifer storage through the introduction of state variables to a groundwater-management formulation. State variables now can be used in the objective function as well as the constraint set of a formulation. Typical groundwater-management applications of state variables are to find withdrawal strategies that meet water-supply demands while simultaneously minimizing drawdowns (Huili and others, 2000; McPhee and Yeh, 2004), streamflow depletions (Male and Mueller, 1992; Mueller and Male, 1993; Eggleston, 2004), or changes in aquifer storage (Bexfield and others, 2004). This new capability is provided by the State Variables (STA) Package for GWM-2005.

This report provides (1) a description of the types of state variables that can be defined in GWM-2005 and the mathematical basis for the use of state variables in GWM-2005, (2) several sample problems that illustrate how state variables can be applied in groundwater-management problems, and (3) instructions for preparing data-input files for a GWM-2005 simulation that includes state variables (appendix 1). Material presented in this report is based on the initial development of the STA Package by Baker (2008).

Detailed background information on the GWM Process is provided by Ahlfeld and others (2005 and 2009), and it is assumed that the reader is familiar with the theory and use of both GWM-2005 and MODFLOW-2005. Because the STA Package is available only for the 2005 version of GWM, the GWM-2005 code is referred to in the remainder of the report by the abbreviated form GWM. Also, throughout the report, all file types are shown in bold uppercase text, such as MODFLOW's **NAME** file.

Description of the State Variables Package

This section provides a description of the types of state variables that can be defined in GWM, the mathematical basis for the use of state variables in GWM, and two extensions of the GWM Process necessary for the use of state variables in a formulation. Details about how the STA Package has been incorporated into the GWM computer code are provided in appendix 2.

Definition of State Variables

In general, a state variable is any simulated condition that defines the state of a groundwater-flow system or a coupled groundwater/surface-water flow system, such as the distribution of hydraulic heads within the flow system, flows into or out of aquifer storage, or streamflow rates within a hydraulically connected stream. The initial version of the STA Package described here allows for the definition and use of three types of state variables in a GWM simulation—heads, streamflows, and changes in aquifer storage, which are referred to simply as storage state variables throughout the report. Regardless of its type, each state variable r is referred to mathematically as S_r.

Head state variables are associated with a single cell in the model domain; streamflow state variables are associated with a single segment-reach location. Each head and streamflow state variable is associated with one and only one stress period, with the value of the state variable taken at the end of the specified stress period. If heads or streamflows at a particular location are to be managed for more than one stress period, multiple state variables must be defined at the location, one for each stress period of interest. Water-level drawdowns and streamflow depletions cannot be specified as state variables directly. Constraints on drawdown or streamflow depletion can be included in a formulation using the existing HEDCON and STRMCON Packages in GWM or by redefining the constraint in terms of head or streamflow state variables. As an example of the latter approach, if there is a requirement to limit drawdown at a particular model cell to 10 ft, and this drawdown would lower the water table at the cell to a minimum of 50 ft above the local datum, an equivalent requirement could be specified with a head type state variable that is constrained to be greater than or equal to 50 ft above the local datum at the cell.

Storage state variables are defined by computing the change in storage over specified regions of the model domain from the beginning to the end of a specified time period. For a single MODFLOW cell and single time step, the change in storage can be computed in a manner similar to that described on pages 5–12 to 5–14 of Harbaugh (2005) and is summarized here for the case in which the cell does not convert between confined and unconfined conditions during the time step:

$$\Delta S = SCB\,(h_{i,j,k}^{m} - TOP_{i,j,k}) + SCA\,(TOP_{i,j,k} - h_{i,j,k}^{m-1})\ , \tag{1}$$

where ΔS is the change in storage (units of length cubed); $h_{i,j,k}^{m}$ is the head at cell i,j,k at the end of time step m (length); $TOP_{i,j,k}$ is the elevation of the top of the model cell (length); and SCA and SCB (length squared) are either the primary or secondary storage capacities depending on confined and unconfined conditions as defined in Harbaugh (2005). Equation 1 is equivalent to equation 5–37 in Harbaugh (2005) but is multiplied by the length of the time step of interest. GWM computes the quantity in equation 1 for each cell in the specified region and for each time step in the specified time period using the same procedures that the MODFLOW Groundwater-Flow (GWF) Process uses for calculation of the volumetric budget. The value assigned to a storage state variable is computed by summing the individual cell values over all cells in the specified region and over all time steps in the specified time period. Although GWM has the capability to simulate groundwater flow by using the MODFLOW Hydrogeologic-Unit Flow (HUF) Package (Anderman and Hill, 2000), the HUF Package cannot be used with storage state variables; therefore, the user must select either the Block-Centered Flow (BCF) or Layer-Property Flow (LPF) Packages described in Harbaugh (2005, chap. 5) to simulate internal groundwater flow with storage state variables.

In GWM, the specified time period for computation of a storage state variable is defined by a beginning and ending stress period. The specified time period extends from the beginning of the first time step in the beginning stress period to the end of the last time step in the ending stress period. The specified region for computation of a storage state variable can include the entire model domain or can be limited to a portion of the domain. When the storage state variable is associated with only a portion of the domain, the specific cells associated with the storage state variable are identified by reading cell-by-cell arrays in the input files.

State variables may be used with the LGR capability of GWM-2005. When using a multimodel simulation (that is, one that includes both a parent model and one or more child models), separate sets of data-input files are used for the parent model and for each child model (see Ahlfeld and others, 2009, for details). Head and (or) streamflow state variables may be placed on the parent grid and (or) any of the child grids.

Storage state variables may be defined over the entirety or just a portion of each individual grid; storage state variables also can be defined that include change in storage in multiple grids. For example, figure 1 illustrates a multigrid model that consists of the parent grid and a single child grid with three different storage state variables. Storage state variable *SSV1* lies entirely within the parent grid and would be described within the GWM input files for that grid. Variable *SSV2* lies entirely within the child grid and would be described using the GWM input files for the child grid. Variable *SSV3* overlies both the parent and child grids. The portions of this state variable that lie within each of the parent and child grids would be described in the corresponding grid input files. The user informs GWM that the parts of *SSV3* defined in the separate input files are associated with the same storage state variable by assigning the same name to the state variable in both input files.

State variables are defined in a GWM simulation by use of a state-variables (**STAVAR**) input file. A **STAVAR** input file must be defined for each parent or child model that includes state variables. Instructions for preparing a **STAVAR** file, as well as modifications to other GWM input files that might be required for a simulation that includes state variables, are provided in appendix 1.

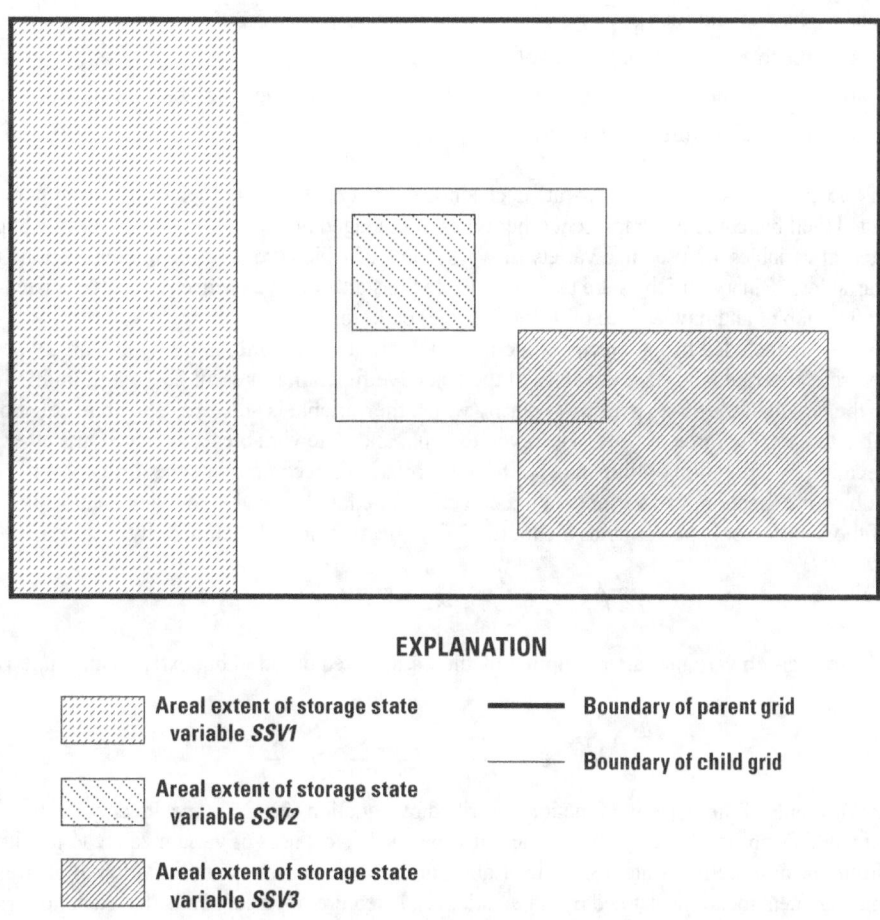

EXPLANATION

Areal extent of storage state variable *SSV1*	——— Boundary of parent grid
	——— Boundary of child grid
Areal extent of storage state variable *SSV2*	
Areal extent of storage state variable *SSV3*	

Figure 1.   Illustration of the use of multiple storage state variables on a multigrid problem. *SSV1, SSV2,* and *SSV3* are regions of the aquifer for which storage state variables are defined. *SSV1* and *SSV2* are on the parent and child grids, respectively, whereas *SSV3* spans both grids.

State Variables in a GWM Formulation

State variables in the STA Package can be used in the objective function and in linear-summation constraints in the same way as flow-rate, external, and binary decision variables available in GWM. The expanded objective function that is now supported by GWM is to minimize or maximize

$$\sum_{n=1}^{N} \beta_n Qw_n T_{Qw_n} + \sum_{m=1}^{M} \gamma_m Ex_m T_{Ex_m} + \sum_{l=1}^{L} k_l I_l + \sum_{r=1}^{R} \xi_r S_r T_{s_r} \, , \tag{2a}$$

where

Qw_n, Ex_m, I_l	are	flow-rate, external, and binary decision variables n, m, and l, respectively;
S_r	is	state variable r;
β_n	is	the cost or benefit per unit volume of water withdrawn or injected at flow-rate decision variable n;
γ_m	is	the unit cost or benefit associated with external variable m;
κ_l	is	the unit cost or benefit associated with binary variable l;
ξ_r	is	the unit cost or benefit associated with state variable r;
T_{Qw_n}	is	the total duration of flow at flow-rate decision variable n;
T_{Ex_m}	is	the total duration of activity of external variable m;
T_{s_r}	is	the total duration of activity of state variable r;
N, M, L	are	the total number of flow-rate, external, and binary decision variables, respectively; and
R	is	the total number of state variables.

As described in Ahlfeld and others (2005), a flow-rate decision variable Qw_n typically represents the withdrawal or injection flow rate at a managed well but could also represent other types of managed flows such as a recharge rate applied to an artificial-recharge basin. External variables are used in a variety of ways, some of which are described in the section "Extended Definitions of External Variables." Binary variables are used to indicate the status (active or inactive) of associated sets of flow-rate and external decision variables and have values of 0 (for inactive) or 1 (for active).

State variables may be included in an objective function that includes any combination of the three types of decision variables. When all flow-rate, external, and state variables of the objective function represent flow rates, multiplication of the value of each variable by the length of the stress periods(s) during which the variable is active, as shown in equation 2a, converts the value of each variable to a total volume of water. However, external and state variables may have units other than flow rate, in which case multiplication by a duration of time may not be appropriate. To accommodate the different types of variables that can be specified in the objective function, two variants of the objective function defined by equation 2a have been added to GWM. In the first, none of the decision or state variables are multiplied by the duration of their activity:

$$\sum_{n=1}^{N} \beta_n Qw_n + \sum_{m=1}^{M} \gamma_m Ex_m + \sum_{l=1}^{L} k_l I_l + \sum_{r=1}^{R} \xi_r S_r \, ; \tag{2b}$$

in the second, flow-rate decision variables are multiplied by their associated duration but external and state variables are not:

$$\sum_{n=1}^{N} \beta_n Qw_n T_{Qw_n} + \sum_{m=1}^{M} \gamma_m Ex_m + \sum_{l=1}^{L} k_l I_l + \sum_{r=1}^{R} \xi_r S_r \, . \tag{2c}$$

One of the three variants of the objective function described by equations 2a–2c is specified by the user in the GWM input files (see "Extended Definitions of Objective-Function Types"). Typical uses of variant 2a are to maximize the volume of water pumped from one or more flow-rate decision variables or to maximize the quantity (volume) of streamflow at one or more streamflow-management locations defined by state variables. A feature of variant 2a is that each flow-rate, external, and state variable in the objective function is weighted by the duration of its activity, so that variables that extend over relatively long periods of time have greater influence on the value of the objective function than those that are shorter in duration. Variant 2b directly maximizes or minimizes rates of flow (that is, volumes per time), such as the rate of pumping at a set of flow-rate decision variables or the rate of streamflow at a set of streamflow-management locations. Variant 2c allows for the option of having different units for the flow-rate decision variables from those for the external and state variables. For example, a user might want to maximize the sum of the volume of water pumped at a set of flow-rate decision variables and the volume of water in a surface-water reservoir as represented by an external variable.

No matter which variant of the objective function is selected, the user must ensure that the units of all terms in the objective function are the same. This requires that the units of the cost/benefit coefficients associated with each decision or state variable (that is, β_n, γ_m, κ_l, and ξ_r) also are selected such that a consistent unit is used for each term of the objective function. The specific units that are selected for the coefficients of the external and state variables will depend on which type(s) of variables have been defined for the formulation (such as heads, flow rates, or storage volumes), as well as the units of the coefficients of the other variables specified in the objective function. Examples of how the units of the objective function can be defined and interpreted are provided in the sample problems described later in the report.

State variables also can be included in the constraint set of a GWM problem using linear-summation constraints (or, simply, summation constraints; Ahlfeld and others, 2005, p. 12–14). These constraints have the general form

$$\sum_{p=1}^{P} a_p GV_p \leq b, \tag{3a}$$

$$\sum_{p=1}^{P} a_p GV_p \geq b, \tag{3b}$$

and

$$\sum_{p=1}^{P} a_p GV_p = b, \tag{3c}$$

where a_p and b are specified coefficients, P is the total number of terms on the left-hand side of the summation, and GV_p is any of the three types of decision variables (Qw_n, Ex_m, I_l) or state variables (S_r) defined for a management problem. Any combination of state variables and flow-rate, external, and binary decision variables may be defined in a linear-summation constraint; however, the generalized nature of these constraints requires that the user ensure that each constraint definition, including the units defined for the coefficients, is logical. Additional details on the use of these constraint types are described in Ahlfeld and others (2005). Examples of the use of state variables in summation constraints are provided in the DEWATER-SV and MAXI-MIN sample problems described later in the report.

State variables can be used in the objective function or summation constraints specified for a multimodel formulation, just as they can be used in a single-model problem that does not use LGR.

Implementation of State Variables in GWM

From the perspective of the groundwater-management formulation, state variables are a type of decision variable. However, the value of each state variable depends on the value of the flow-rate decision variables as determined by the flow-process simulation. Hence, state variables can be considered secondary (or derived) decision variables, in contrast to the flow-rate variables, which are the primary decision variables of the groundwater-management formulation. The STA Package takes advantage of the relation between the flow-rate and state variables by substituting all occurrences of state variables in the formulation with a first-order Taylor series approximation written in terms of the flow-rate variables. This substitution is done automatically by the STA Package. From the user's perspective, a formulation that includes state variables can be conveniently described; the input to GWM and the STA Package is provided in terms of state variables, and the output from GWM reports the value of the state variables. However, within the GWM solution algorithm, state variables are converted to their equivalent Taylor-series approximation.

The use of the first-order Taylor-series expansion for groundwater-management formulations is described in detail in Ahlfeld and Mulligan (2000, p. 64–66) and Ahlfeld and others (2005, p. 20). In the STA Package for a given state variable, the Taylor-series expansion is defined as

$$S_r(Qw) = S^o_r(Qw^o) + \sum_{n=1}^{N} \frac{\partial S_r}{\partial Qw_n}(Qw^o)(Qw_n - Qw^o_n), \tag{4}$$

where

$S_r(Qw)$ is the value of the state variable for a new vector (that is, a new set) of withdrawal and injection flow rates Qw having individual elements Qw_n;

$S^o_r(Qw^o)$ is the value of the state variable for an original vector (that is, a base-condition set) of withdrawal and injection flow rates Qw^o having individual elements Qw^o_n;

$\dfrac{\partial S_r}{\partial Q w_n}(\boldsymbol{Q w^o})$ is   the change in the value of the state variable for a change in withdrawal or injection flow rate for the nth flow-rate decision variable, evaluated at the original vector of flow rates $\boldsymbol{Q w^o}$; and

N is   the total number of flow-rate decision variables.

The partial derivatives in equation 4, $\dfrac{\partial S_r}{\partial Q w_n}$, are called the response coefficients. Each response coefficient quantifies the change in the value of the state variable with a change in the withdrawal or injection rate at decision variable $Q w_n$. Response coefficients are calculated by GWM by use of the perturbation technique described in Ahlfeld and others (2005, p. 21).

The right-hand side of equation 4 is substituted for each state variable in the objective function (eqs. 2a–2c) and in each summation constraint (eqs. 3a–3c) as needed. An expansion of the right-hand side of equation 4 results in three separate terms that must be considered in the substitution of each state variable, specifically

$$S_r(\boldsymbol{Q w}) = S^o_r(\boldsymbol{Q w^o}) + \sum_{n\,1}^{N} \frac{\partial S_r}{\partial Q w_n}(\boldsymbol{Q w^o})(Q w_n) - \sum_{n\,1}^{N} \frac{\partial S_r}{\partial Q w_n}(\boldsymbol{Q w^o})(Q w^o_n) \cdot \tag{5}$$

The first and third terms are constants because the values of the response coefficients, the base-condition state variables ($S^o_r(\boldsymbol{Q w^o})$), and the base-condition flow rates ($Q w_n^o$) are known when the Taylor-series approximation is constructed; only the values of the optimal flow rates, $Q w_n$, are unknown. For the case of a summation constraint, all three terms on the right-hand side of equation 5 must be retained in the substitution of each state variable. However, because constants in the objective function do not affect the solution of a management formulation, the only term on the right-hand side of equation 5 that is retained

for each state variable in the objective function is $\displaystyle\sum_{n=1}^{N} \frac{\partial S_r}{\partial Q w_n}(\boldsymbol{Q w^o})(Q w_n)$ to obtain

$$S_r(\boldsymbol{Q w}) = \sum_{n=1}^{N} \frac{\partial S_r}{\partial Q w_n}(\boldsymbol{Q w^o})(Q w_n) \cdot \tag{6}$$

After the optimal flow rates have been determined, the two constant terms in equation 5 are added back into the equation to determine the total value of the objective function.

As an example of the use of equation 6, assume that the objective of a particular management problem is to maximize head at a particular location in a simulated aquifer. This head has been represented by state variable $H1$ and the objective is therefore to

$$\textit{Maximize } H1 \;. \tag{7}$$

Further assume that three flow-rate decision variables that represent withdrawal rates at three managed wells have been defined for the problem, $Q1$, $Q2$, and $Q3$. On the basis of equation 6, $H1$ is therefore replaced in the objective function by

$$\frac{\partial H1}{\partial Q1}Q1 + \frac{\partial H1}{\partial Q2}Q2 + \frac{\partial H1}{\partial Q3}Q3 \;, \tag{8}$$

where the partial derivatives $\dfrac{\partial H1}{\partial Q1}$, $\dfrac{\partial H1}{\partial Q2}$, and $\dfrac{\partial H1}{\partial Q3}$ are response coefficients calculated by GWM.

Similarly, assume that a constraint has been defined to ensure that head at $H1$ must be greater than or equal to a specified minimum value $H1^*$: $H1 \geq H1^*$. In this case, $H1$ is defined by equation 5 as

$$H1 = H1^o + \left[\frac{\partial H1}{\partial Q1}Q1 + \frac{\partial H1}{\partial Q2}Q2 + \frac{\partial H1}{\partial Q3}Q3\right] - \left[\frac{\partial H1}{\partial Q1}Q1^o + \frac{\partial H1}{\partial Q2}Q2^o + \frac{\partial H1}{\partial Q3}Q3^o\right] \;, \tag{9}$$

where $H1^o$ is the base-condition head at $H1$ and $Q1^o$, $Q2^o$, and $Q3^o$ are the base-condition flow rates. The resulting constraint is

$$H1^o + \left[\frac{\partial H1}{\partial Q1}Q1 + \frac{\partial H1}{\partial Q2}Q2 + \frac{\partial H1}{\partial Q3}Q3\right] - \left[\frac{\partial H1}{\partial Q1}Q1^o + \frac{\partial H1}{\partial Q2}Q2^o + \frac{\partial H1}{\partial Q3}Q3^o\right] \geq H1^* \;. \tag{10}$$

The substitutions described by equations 4 through 6 are done automatically by GWM as part of the solution process. That is, the user defines the state variables for the management problem and GWM then makes the conversions from state variables to flow-rate decision variables in the objective function and summation constraints as necessary.

Additional Extensions to GWM-2005 Necessary for the State Variables Package

The addition of state variables to GWM necessitated changes to the existing functionality of the external decision variables and objective-function types defined for a GWM problem. These changes are described below.

Extended Definitions of External Variables

External decision variables were developed for the initial releases of GWM as general-purpose variables that do not have a direct effect on the state variables of the groundwater-flow system (heads, streamflows, and so forth). Two types of external variables were defined, imports (or sources) of water and exports (or sinks) of water. A typical use of an external variable is to represent water that is imported to a groundwater basin from an out-of-basin surface-water reservoir. This water might be used, for example, to supplement water-supply demands that cannot be met by withdrawals from aquifers within a basin. External variables can be used in the objective function or as part of summation constraints of a management problem.

With the addition of state variables to GWM, there are other potential uses of these general-purpose variables. In order to account for these potential uses and to make the output that is generated as part of a GWM simulation as informative as possible, four new types of external decision variables are provided in GWM in addition to the import (IM) and export (EX) types previously available. These new types are head (HD), streamflow (SF), storage (ST), and general (GN). A common use of these types of external variables is in management formulations named minimax and maximin that are described in detail for the MAXIMIN sample problem. Examples of minimax and maximin formulations are to "minimize the maximum storage change throughout a basin" or "maximize the minimum head at a set of water-level control points." In the first example, a storage-type external variable is defined to represent the maximum storage change throughout the basin, and in the second, a head-type external variable is defined to represent the minimum head. In each case, the optimal value of the external variable is determined as part of the problem solution. The GN external-variable type has been added to provide additional flexibility to formulate management problems with GWM. For example, the GN external variable might be used to represent flows and storage of water that are not part of the simulated groundwater system, such as canal or river flows or storage in a surface-water reservoir.

The type of external variable is specified by use of input variable ETYPE in the decision variables (**DECVAR**) input file. Regardless of the type of external variable defined, all external variables are treated as positive-valued variables. As a result, the selection of the sign of the coefficients for the external variables in the objective function (eqs. 2a–2c) and summation constraints (eqs. 3a–3c) is important in properly using these variables. Also, as with state variables, the units that are associated with each external variable, such as length for a HD type or cubic length per time for a SF type, must be carefully considered.

Extended Definitions of Objective-Function Types

As described by equations 2a–2c, GWM now supports three variants of the objective function. The variant selected for a particular management formulation is specified in the **OBJFNC** input file by use of variable FNTYP in item 2. FNTYP must be specified as one of the following:

WSDV—flow-rate, external, and state variables are automatically multipled by the duration of their activity (eq. 2a).

USDV—none of the decision or state variables are multiplied by the duration of their activity (eq. 2b).

MSDV—flow-rate decision variables are multiplied by their associated duration, but external and state variables are not (eq. 2c).

Sample Problems

Four sample problems are provided to illustrate how state variables can be used in a groundwater-management formulation. Sample problems are provided for each of the three types of state variables described in this report: heads (sample problems DEWATER-SV and MAXIMIN), streamflows (STREAMFLOW), and change in aquifer storage (STORAGE). Selected input and output files are included with each of the sample problems; all input files for each of the sample problems are included in the GWM-2005 distribution package available at the U.S. Geological Survey web site provided in the Preface to this report.

DEWATER-SV: Minimize Cost of Lowering Groundwater Levels

This sample problem represents a steady-state dewatering problem for a construction site and is based on the DEWATER problem presented in Ahlfeld and others (2005). The objective of the groundwater-management problem is to minimize the cost of lowering groundwater levels at the construction site so that footings can be installed in the area shown in figure 2. The purpose of the sample problem is to demonstrate that head-type state variables can be used interchangeably with head-type constraints in a management formulation; this is possible because both approaches for representing heads in a management formulation are based on the relation between heads and flow-rate decision variables described by equation 5.

The aquifer at the site is confined and is simulated by a single model layer that is 3,000 ft long and 2,000 ft wide. The model grid consists of 20 rows and 30 columns, and each grid cell is 100 ft by 100 ft (fig. 2). The model uses no-flow boundary conditions along the northern and southern boundaries of the aquifer and constant heads of 80 ft and 60 ft along the western and eastern boundaries of the aquifer, respectively. The transmissivity of the aquifer is 50 ft²/d.

The MODFLOW model consists of a **NAME** file, a Discretization (**DIS**) file, a Basic (**BAS**) file, a Block-Centered Flow (**BCF**) file, and a Preconditioned Conjugate-Gradient (**PCG**) file.

Groundwater-Management Problem

In this example formulation, only the operational costs of pumping groundwater to lower water levels at the site are considered. Seven candidate well locations are selected as possible locations of withdrawal (fig. 2), and the objective of the management problem is to minimize the cost of pumping water from the seven candidate wells over a period of 1,000 days:

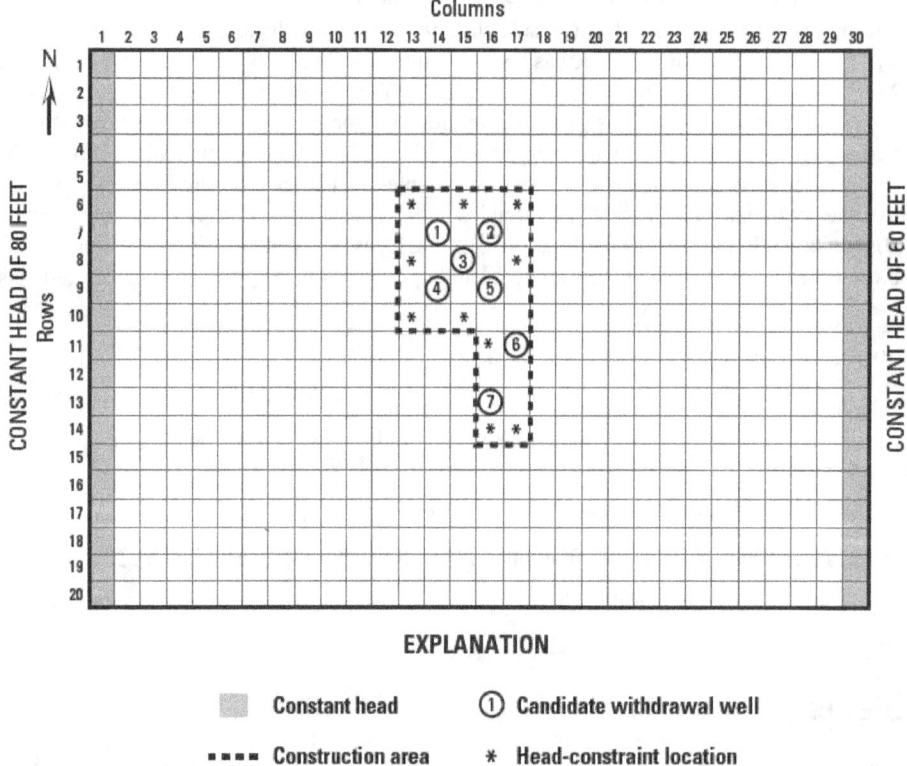

EXPLANATION

Constant head   ① Candidate withdrawal well

■ ■ ■ ■ Construction area   * Head-constraint location

Figure 2.   Schematic diagram showing model grid for DEWATER-SV sample problem. The head-constraint locations coincide with the locations of head-type state variables.

$$\text{Minimize} \sum_{n=1}^{7} Q_n \ , \tag{11}$$

where Q_n is the pumping rate at well n. The decision variables are named $Q1$, $Q2$, and so forth. The minimum and maximum pumping rates at each well are 0 and 20,000 ft³/d, respectively. The decision variable (**DECVAR**), decision-variable constraint (**VARCON**), and objective function (**OBJFNC**) files for the sample problem are unchanged from the original DEWATER problem described in Ahlfeld and others (2005).

Groundwater levels at the construction site are to be lowered to a maximum elevation of 50 ft at each of the 10 constraint locations shown on figure 2. Each of the water-level constraints can be written as

$$h_{j,i,k} \leq 50 \ , \tag{12}$$

where $h_{j,i,k}$ is the optimal water level (head) at location i, j, k. In the original sample problem, these constraints were specified in a head-constraints (**HEDCON**) file. Here, the heads at the 10 constraint locations are defined as state variables in a **STAVAR** file and the constraints themselves are specified in a summation constraints (**SUMCON**) file. Each state variable is named b-01, b-02, and so forth.

Because the aquifer is confined and there are no simulated head-dependent boundary conditions, the management problem is linear and is solved by use of the LP (linear programming) option in the solution and output control file (**SOLN**) file. The GWM input files for the formulation are listed at the end of the sample problem.

The response coefficient calculated by GWM for each state-variable/decision-variable pair is exactly equal to the response coefficient calculated for each head-constraint/decision-variable pair in the original problem; the full response matrix for this sample problem is shown in figure 3.

Figure 3. Response matrix calculated by GWM for the DEWATER-SV sample problem.

```
STATE VARIABLE RESPONSE MATRIX

  ROWS FOR STATE VARIABLES IN READ-ORDER FROM STAVAR
  COLUMNS IN READ-ORDER FOR FLOW RATE VARIABLES
-----------------------------------------------------------


                    1              2              3              4              5
                    6              7
.....................................................................
   1    0.101949E-01   0.778914E-02   0.797034E-02   0.750478E-02   0.660235E-02
        0.517529E-02   0.463308E-02
   2    0.103054E-01   0.103421E-01   0.929592E-02   0.761145E-02   0.764697E-02
        0.579649E-02   0.494582E-02
   3    0.782590E-02   0.103054E-01   0.804271E-02   0.663787E-02   0.761145E-02
        0.599856E-02   0.491298E-02
   4    0.991186E-02   0.752346E-02   0.894489E-02   0.972287E-02   0.734398E-02
        0.579601E-02   0.534809E-02
   5    0.755933E-02   0.100196E-01   0.901600E-02   0.737916E-02   0.982850E-02
        0.747409E-02   0.582287E-02
   6    0.731580E-02   0.642287E-02   0.759578E-02   0.961431E-02   0.724018E-02
        0.636208E-02   0.631907E-02
   7    0.742033E-02   0.745515E-02   0.890618E-02   0.971859E-02   0.975333E-02
        0.832910E-02   0.734480E-02
   8    0.628342E-02   0.668314E-02   0.734480E-02   0.759432E-02   0.883312E-02
        0.110537E-01   0.890618E-02
   9    0.479245E-02   0.493955E-02   0.534692E-02   0.568859E-02   0.595585E-02
        0.742033E-02   0.114198E-01
  10    0.463179E-02   0.487710E-02   0.519641E-02   0.541579E-02   0.585875E-02
        0.758425E-02   0.100196E-01
```

The solution to the management formulation is exactly the same as that determined for the equivalent formulation described in Ahlfeld and others (2005). The value of the objective function at the optimal solution is 2.8657×10^6 ft^3 of water withdrawn (as shown in the GWM output file listed at the end of the sample problem). Four wells were selected for pumping in the optimal solution: well $Q1$ (1,077.4 ft^3/d), well $Q2$ (78.2 ft^3/d), well $Q4$ (769.0 ft^3/d), and well $Q7$ (941.1 ft^3/d). Four of the summation constraints are binding at the solution, those associated with state variables b-01, b-03, b-06, and b-10. The value of each state variable is shown as part of the optimal-solution report in the GWM output file. Because none of the state variables are included in the objective function of this formulation, they do not contribute to the optimal value of the objective function. The **OUT** file lists the values of all 10 state variables computed using the state-variable response matrix and by running the GWF Process with the optimal pumping rates. As expected, the two sets of heads are identical.

Selected Input and Output Files

NAME file (*dewatersv.nam*)

```
LIST   10    dewatersv.lst
DIS    11    ..\data\DEWATER-SV\dewatersv.dis
BAS6   12    ..\data\DEWATER-SV\dewatersv.ba6
BCF6   13    ..\data\DEWATER-SV\dewatersv.bc6
PCG    14    ..\data\DEWATER-SV\dewatersv.pcg
GWM    15    ..\data\DEWATER-SV\dewatersv.gwm
```

GWM file (*dewatersv.gwm*)

```
#DEWATER-SV Sample Problem, GWM file
#December 2009
OUT      dewatersv.gwmout
DECVAR   ..\data\DEWATER-SV\dewatersv.decvar
STAVAR   ..\data\DEWATER-SV\dewatersv.stavar
OBJFNC   ..\data\DEWATER-SV\dewatersv.objfnc
VARCON   ..\data\DEWATER-SV\dewatersv.varcon
SUMCON   ..\data\DEWATER-SV\dewatersv.sumcon
SOLN     ..\data\DEWATER-SV\dewatersv.soln
```

Decision variable (DECVAR) file (*dewatersv.decvar*)

```
#DEWATER-SV Sample Problem, DECVAR file
#December 2009
 1 0                        #1-IPRN  GWMWFILE
 7 0 0                      #2-NFVAR   NEVAR   NBVAR
 Q1  1   1  7  14  W  Y  1  #3a-FVNAME NC LAY ROW COL FTYPE FSTAT WSP
 Q2  1   1  7  16  W  Y  1
 Q3  1   1  8  15  W  Y  1
 Q4  1   1  9  14  W  Y  1
 Q5  1   1  9  16  W  Y  1
 Q6  1   1 11  17  W  Y  1
 Q7  1   1 13  16  W  Y  1
```

State variables file (STAVAR) file (*dewatersv.stavar*)

```
#DEWATER-SV Sample Problem, STAVAR file
#December 2009
1                        #1-IPRN
10  0  0                 #2-NHVAR NRVAR  NSVAR
b-01  1  6 13 1          #3-SVNAME LAY ROW COL SVSP
b-02  1  6 15 1
b-03  1  6 17 1
```

```
b-04  1   8  13  1
b-05  1   8  17  1
b-06  1  10  13  1
b-07  1  10  15  1
b-08  1  11  16  1
b-09  1  14  16  1
b-10  1  14  17  1
```

Objective function (OBJFNC) file (*dewatersv.objfnc*)

```
#DEWATER-SV Sample Problem, OBJFNC file
#December 2009
 1                  #1-IPRN
 MIN  WSDV          #2-OBJTYP  FNTYP
 7  0   0   0       #3-NFVOBJ  NEVOBJ  NBVOBJ  NSVOBJ
 Q1  1.0            #4-FVNAME  FVOBJC
 Q2  1.0
 Q3  1.0
 Q4  1.0
 Q5  1.0
 Q6  1.0
 Q7  1.0
```

Decision-variable constraints (VARCON) file (*dewatersv.varcon*)

```
#DEWATER-SV Sample Problem, VARCON file
#December 2009
 1                          #1-IPRN
 Q1 0.0d2  2.0d4  0.0d2     #2-FVNAME  FVMIN  FVMAX  FVREF
 Q2 0.0d2  2.0d4  0.0d2
 Q3 0.0d2  2.0d4  0.0d2
 Q4 0.0d2  2.0d4  0.0d2
 Q5 0.0d2  2.0d4  0.0d2
 Q6 0.0d2  2.0d4  0.0d2
 Q7 0.0d2  2.0d4  0.0d2
```

Linear-summation constraints (SUMCON) file (*dewatersv.sumcon*)

```
#DEWATER-SV Sample Problem, SUMCON file
#December 2009
 1                          #1-IPRN
 10                         #2-SMCNUM
 Hedcon1  1 le 50.0         #3a-SMCNAME NTERMS TYPE RHS
  b-01 1.0                  #3b-GVNAME  GVCOEFF
 Hedcon2  1 le 50.0
  b-02 1.0
 Hedcon3  1 le 50.0
  b-03 1.0
 Hedcon4  1 le 50.0
  b-04 1.0
 Hedcon5  1 le 50.0
  b-05 1.0
 Hedcon6  1 le 50.0
  b-06 1.0
 Hedcon7  1 le 50.0
  b-07 1.0
```

```
 Hedcon8    1 le 50.0
   b-08 1.0
 Hedcon9    1 le 50.0
   b-09 1.0
 Hedcon10    1 le 50.0
   b-10 1.0
```

Solution and output control (SOLN) file (*dewatersv.soln*)

```
#DEWATER-SV Sample Problem, SOLN file
#December 2009
 LP               #1-SOLNTYP
 3                #4a-IRM
 1000   2000      #4b-LPITMAX  BBITMAX
 0.5              #4c-DELTA
 1  10  0.5  0.0  #4d-NSIGDIG  NPGNMX  PGFACT  CRITMFC
 1   0            #4e-BBITPRT  RANGE
dewater.resp      #4f-RMNAME1
 0                #6a-IBASE
```

Part of the GWM-2005 output file (*dewatersv.gwmout*)

```
-----------------------------------------------------------------------
                 Groundwater Management Solution
-----------------------------------------------------------------------


       OPTIMAL SOLUTION FOUND

       OPTIMAL RATES FOR EACH FLOW VARIABLE
       -----------------------------------------

Variable         Withdrawal       Injection        Contribution
Name             Rate             Rate             To Objective
----------       --------------   ------------     ------------
  Q1             1.077390E+03                      1.077390E+06
  Q2             7.823877E+01                      7.823877E+04
  Q3             0.000000E+00                      0.000000E+00
  Q4             7.689506E+02                      7.689506E+05
  Q5             0.000000E+00                      0.000000E+00
  Q6             0.000000E+00                      0.000000E+00
  Q7             9.410751E+02                      9.410751E+05
                 --------------   ------------     ------------
TOTALS           2.865655E+03     0.000000E+00     2.865655E+06

       OPTIMAL VALUES FOR EACH STATE VARIABLE
       -----------------------------------------

Variable                                          Contribution
Name             Value                            To Objective
----------       ------------                     ------------
  b-01           5.000000E+01                      0.000000E+00
  b-02           4.792554E+01                      0.000000E+00
  b-03           5.000000E+01                      0.000000E+00
  b-04           4.794721E+01                      0.000000E+00
  b-05           4.888328E+01                      0.000000E+00
  b-06           5.000000E+01                      0.000000E+00
  b-07           4.738183E+01                      0.000000E+00
  b-08           4.814155E+01                      0.000000E+00
```

```
b-09              4.898424E+01                      0.000000E+00
b-10              5.000000E+01                      0.000000E+00
                  -------------                     ------------
TOTALS            4.892637E+02                      0.000000E+00


      OBJECTIVE FUNCTION VALUE                2.865655E+06

      BINDING CONSTRAINTS
Constraint Type       Name       Status      Shadow Price
---------------       ----       ------      ------------
Summation             Hedcon1    Binding     -2.7273E+04
Summation             Hedcon3    Binding     -3.2593E+04
Summation             Hedcon6    Binding     -3.1185E+04
Summation             Hedcon10   Binding     -5.1544E+04
```

Binding constraint values are determined from the linear program
   and based on the response matrix approximation of the flow process.

      Range Analysis Not Reported

            Final Flow Process Simulation

Running Final Flow Process Simulation
   using Optimal Flow Variable Rates

      Status of State Variable Values
         Using Optimal Flow Rate Variable Values
      State Variable Type   Name   Computed Value
      -------------------   ----   --------------
      Head   b-01   5.0000000E+01
      Head   b-02   4.7925541E+01
      Head   b-03   5.0000000E+01
      Head   b-04   4.7947208E+01
      Head   b-05   4.8883277E+01
      Head   b-06   5.0000000E+01
      Head   b-07   4.7381829E+01
      Head   b-08   4.8141550E+01
      Head   b-09   4.8984245E+01
      Head   b-10   5.0000000E+01
```

Precision limitations and nonlinear response may cause
   the state variables computed directly by the flow process
   to differ from those computed using the linear program.

      Status of Simulation-Based Constraints
         Using Optimal Flow Rate Variable Values

```
                                    Simulated   Specified
                                    By Flow   in
Constraint Type   Name   Process   Constraints   Difference
---------------   ----   ----------   -----------   ----------
Summation   Hedcon1   5.0000E+01 < 5.0000E+01   -3.4739E-07
Summation   Hedcon2   4.7926E+01 < 5.0000E+01   -2.0745E+00
Summation   Hedcon3   5.0000E+01 < 5.0000E+01   -2.8216E-07
Summation   Hedcon4   4.7947E+01 < 5.0000E+01   -2.0528E+00
Summation   Hedcon5   4.8883E+01 < 5.0000E+01   -1.1167E+00
```
```

```
Summation              Hedcon6      5.0000E+01 < 5.0000E+01   -2.6732E-07
Summation              Hedcon7      4.7382E+01 < 5.0000E+01   -2.6182E+00
Summation              Hedcon8      4.8142E+01 < 5.0000E+01   -1.8584E+00
Summation              Hedcon9      4.8984E+01 < 5.0000E+01   -1.0158E+00
Summation              Hedcon10     5.0000E+01 < 5.0000E+01   -1.8573E-07
```

```
Difference is computed by subtracting right hand side of the constraint
  from the left side of the constraint.
Precision limitations and nonlinear response may cause the
  values of the binding constraints computed directly by the flow process
  to differ from those computed using the linear program.
```

MAXIMIN: Maximize the Minimum Hydraulic Head

This sample problem illustrates how state variables can be used in a common type of resource-management problem that is often referred to as a maximin or minimax formulation. Minimax and maximin formulations are used when the goals of the management problem are to make as much progress as possible toward meeting broad, or open ended, objectives (as described by Hillier and Lieberman, 1990, p. 277–279). These types of formulations are described first through a hypothetical example using streamflow state variables to provide background information on maximin and minimax formulations and then by an example GWM application in which head-type state variables are used to control land subsidence.

Maximin and Minimax Formulations in Groundwater Management

It is sometimes not possible, or not desirable, to explicitly define the goals of each component of a management formulation. For example, a resource manager may have a general goal of maximizing groundwater levels as much as possible throughout a watershed for a specified level of water-supply development, instead of requiring that groundwater levels be greater than or equal to specific values at a set of water-level constraint locations. Objectives such as these can be addressed through the use of minimax and maximin management formulations; that is, formulations in which the goal is to either minimize the maximum value of the management objective or to maximize the minimum value of the objective. Examples of these types of formulations are to "minimize the maximum drawdown throughout a basin" or "maximize the minimum summertime streamflow at a set of streamflow control points." Minimax and maximin formulations require the introduction of a new type of variable to a management formulation; these variables have been referred to as auxiliary variables (Hillier and Lieberman, 1990, p. 278). In GWM, auxiliary variables are defined by use of external variables, and they are often used in conjunction with state variables.

An example groundwater-management problem illustrates the use of an auxiliary variable in a maximin formulation. In this example, the goal is to maximize the minimum streamflow at a set of four streamflow-constraint locations. The four streamflow rates are managed by use of state variables defined at each of the streamflow-management locations $SF1$, $SF2$, $SF3$, and $SF4$. Simultaneously, a minimum withdrawal rate equal to D must be obtained from a set of three candidate pumping wells, defined as flow-rate decision variables $Q1$, $Q2$, and $Q3$. An external (auxiliary) variable R is defined as the minimum streamflow at the four streamflow-management locations, and the value of R is to be determined as part of the problem solution. The objective of the formulation is

$$\text{Maximize } R \text{ ,} \tag{13}$$

subject to the set of constraints

$$SF1 \geq R \text{ ,} \tag{14a}$$

$$SF2 \geq R \text{ ,} \tag{14b}$$

$$SF3 \geq R \text{ ,} \tag{14c}$$

$$SF4 \geq R \text{ , and} \tag{14d}$$

$$Q1 + Q2 + Q3 \geq D \text{ .} \tag{14e}$$

The formulation is designed to make R as large as possible while simultaneously ensuring that the minimum streamflow value at the four streamflow-management locations is at least as large as R. In effect, R defines the minimum value of streamflow at each of the four locations.

Groundwater-Management Problem

This example application illustrates the use of head-type state variables to control land subsidence within a portion of an aquifer that is used for water supply (fig. 4). Although there are alternative ways to formulate a management problem to control land subsidence, the approach used in this sample problem maximizes the minimum groundwater level (hydraulic head) at a set of water-level control points; that is, a maximin formulation is used. The groundwater system consists of a single-layer, confined aquifer that is 3,000 ft long and 2,000 ft wide. The aquifer has a uniform thickness of 100 ft and a uniform, isotropic horizontal hydraulic conductivity of 0.5 ft/d; the transmissivity of the aquifer is therefore 50 ft²/d. The model grid used to represent the aquifer consists of 20 rows and 30 columns, and each grid cell is 100 ft on each side (fig. 4). The model uses no-flow boundary conditions along the northern and southern boundaries of the aquifer and constant heads of 80 ft and 60 ft along the western and eastern boundaries of the aquifer, respectively. The system is managed for steady-state conditions, and a single stress period with an arbitrary length of 1,000 days is used in the simulation.

The MODFLOW model consists of a **NAME** file, a Discretization (**DIS**) file, a Basic (**BAS**) file, a Layer-Property Flow (**LPF**) file, and a Preconditioned Conjugate-Gradient (**PCG**) file.

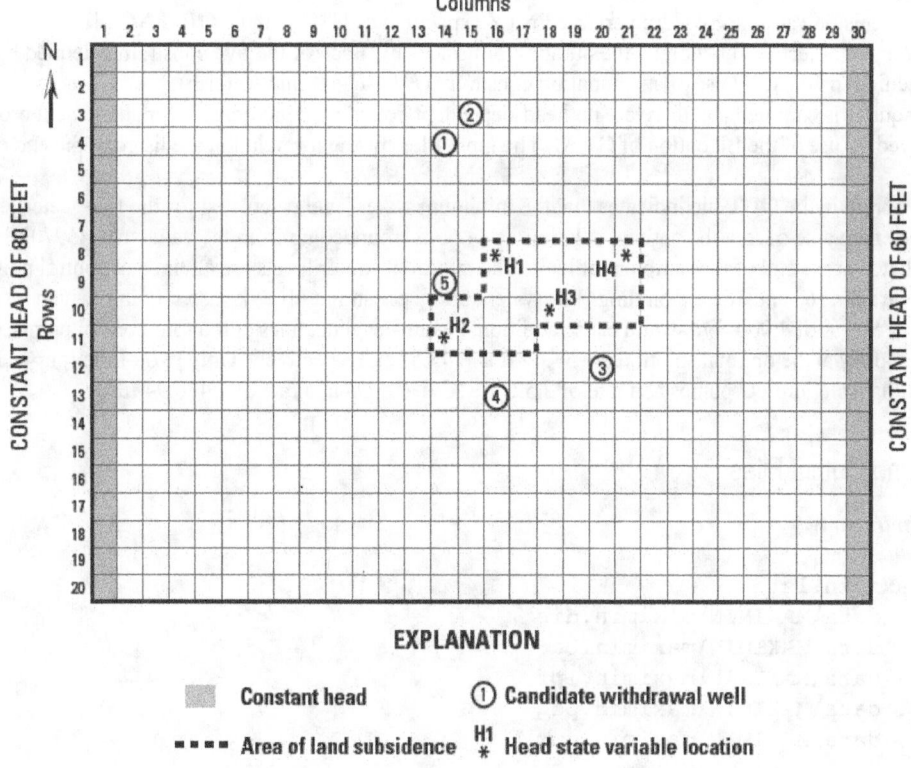

Figure 4. Schematic diagram showing model grid for MAXIMIN sample problem.

The decision variables specified for the management problem are withdrawal rates at five candidate well locations whose locations are shown in figure 4 (wells $Q1$, $Q2$, and so forth). The minimum and maximum withdrawal rates specified for each of the five wells are 0 and 4,000 ft³/d, respectively. A head-type external decision variable (R), which represents the minimum water level at the four control points, also is defined for the problem. Bounds on R are set at 0 and 1,000 ft, respectively—values that are not expected to be reached in the solution. The unknown values of head at each of the four control points are represented by four head-type state variables, $H1$, $H2$, $H3$, and $H4$. A total withdrawal rate of 4,000 ft³/d is sought from the five candidate wells to meet water-supply demands.

The problem is formulated as a maximin problem, with the objective of finding the maximum value of R that makes the minimum water level at the four control points as large as possible:

$$\text{Maximize } R \text{ ,} \tag{15}$$

subject to the set of constraints

$$H1 \geq R \text{ ,} \tag{16a}$$

$$H2 \geq R \text{ ,} \tag{16b}$$

$$H3 \geq R \text{ ,} \tag{16c}$$

$$H4 \geq R \text{ ,} \tag{16d}$$

$$Q1 + Q2 + Q3 + Q4 + Q5 \geq 4,000 \text{ , and} \tag{16e}$$

the upper and lower limits on each of the decision variables described previously. The objective-function type is defined as an unweighted sum of decision variables (eq. 2b; variable FNTYP specified as USDV in the **OBJFNC** file) because there is no need to multiply R in the objective function by the duration of its activity. Each of the five constraints defined by equations 16a through 16e is specified in GWM by use of a summation constraint ($H1 - R \geq 0$, and so forth).

Because the aquifer is confined and there are no head-dependent boundary conditions, the management problem is linear and can be solved by use of the LP option of GWM. The input files for the formulation are listed at the end of the sample problem.

The output reported in the **OUT** file indicates that the minimum water level calculated for the four water-level control points, which is also the value of R at the optimal solution, is 44.60 ft (rounded up from the value of 4.459702E+01 shown in the GWM output file listed at the end of the sample problem). The minimum water levels are attained at control points $H1$ and $H3$, and, therefore, constraints 16a and 16c are binding. The water-supply constraint (16e) is also binding, which results from the fact that any pumping in excess of 4,000 ft³/d would cause additional water-level declines at the four control points. Water levels at control points $H2$ and $H4$ at the optimal solution are 51.07 ft and 45.58 ft, respectively. Only two of the four wells were selected to pump in the optimal solution; $Q2$ pumps at a rate of 2,523.06 ft³/d and $Q3$ at a rate of 1,476.94 ft³/d.

Selected Input and Output Files

NAME file (*maximin.nam*)

```
LIST   10    maximin.lst
DIS    11    ..\data\MAXIMIN\maximin.dis
BAS6   12    ..\data\MAXIMIN\maximin.ba6
LPF    13    ..\data\MAXIMIN\maximin.lpf
PCG    14    ..\data\MAXIMIN\maximin.pcg
GWM    15    ..\data\MAXIMIN\maximin.gwm
```

GWM file (*maximin.gwm*)

```
#MAXIMIN Sample Problem, GWM file
#December 2009
OUT    maximin.gwmout
```

```
DECVAR    ..\data\MAXIMIN\maximin.decvar
STAVAR    ..\data\MAXIMIN\maximin.stavar
OBJFNC    ..\data\MAXIMIN\maximin.objfnc
VARCON    ..\data\MAXIMIN\maximin.varcon
SUMCON    ..\data\MAXIMIN\maximin.sumcon
SOLN      ..\data\MAXIMIN\maximin.soln
```

Decision variable (DECVAR) file (*maximin.decvar*)

```
#MAXIMIN Sample Problem, DECVAR file
#December 2009
 1 0                              #1-IPRN   GWMWFILE
 5 1 0                            #2-NFVAR   NEVAR   NBVAR
Q1  1   1  4  14   W  Y  1        #3a-FVNAME NC LAY ROW COL FTYPE FSTAT WSP
Q2  1   1  3  15   W  Y  1
Q3  1   1 12  20   W  Y  1
Q4  1   1 13  16   W  Y  1
Q5  1   1  9  14   W  Y  1
R   HD  1                         #4 EVNAME ETYPE ESP
```

State variables file (STAVAR) file (*maximin.stavar*)

```
#MAXIMIN Sample Problem, STAVAR file
#December 2009
 1                                #1-IPRN
 4  0  0                          #2-NHVAR NRVAR NSVAR
H1  1   8  16   1                 #3-SVNAME LAY ROW COL SVSP
H2  1  11  14   1
H3  1  10  18   1
H4  1   8  21   1
```

Objective function (OBJFNC) file (*maximin.objfnc*)

```
#MAXIMIN Sample Problem, OBJFNC file
#December 2009
 1              #1-IPRN
MAX  USDV       #2-OBJTYP  FNTYP
 0  1  0  0     #3-NFVOBJ   NEVOBJ   NBVOBJ  (NSVOBJ)
R  1.0          #5-EVNAME  EVOBJC
```

Decision-variable constraints (VARCON) file (*maximin.varcon*)

```
#MAXIMIN Sample Problem, VARCON file
#December 2009
 1                            #1-IPRN
Q1 0.0d2  4.0d3  0.0d2        #2-FVNAME FVMIN  FVMAX  FVREF
Q2 0.0d2  4.0d3  0.0d2
Q3 0.0d2  4.0d3  0.0d2
Q4 0.0d2  4.0d3  0.0d2
Q5 0.0d2  4.0d3  0.0d2
R  0.0d0  1.0d3               #3-EVNAME  EVMIN  EVMAX
```

Linear-summation constraints (SUMCON) file (*maximin.sumcon*)

```
#MAXIMIN Sample Problem, SUMCON file
#December 2009
```

```
1                        #1-IPRN
5                        #2-SMCNUM
DEMAND 5 ge 4000.        #3a-SMCNAME NTERMS TYPE RHS
 Q1  1.0                 #3b-GVNAME GVCOEFF
 Q2  1.0
 Q3  1.0
 Q4  1.0
 Q5  1.0
CON1 2 ge 0.0
 H1  1.0
 R   -1.0
CON2 2 ge 0.0
 H2  1.0
 R   -1.0
CON3 2 ge 0.0
 H3  1.0
 R   -1.0
CON4 2 ge 0.0
 H4  1.0
 R   -1.0
```

Solution and output control (SOLN) file (*maximin.soln*)

```
#MAXIMIN Sample Problem, SOLN file
#December 2009
 LP                     #1-SOLNTYP
 3                      #4a-IRM
 1000   2000            #4b-LPITMAX   BBITMAX
 0.5                    #4c-DELTA
 1  10   0.5            #4d-NSIGDIG   NPGNMX   PGFACT
 1   0                  #4e-BBITPRT   RANGE
maximin.resp            #4f-RMNAME1
 0                      #6a-IBASE
```

Part of the GWM-2005 output file (*maximin.out*)

```
------------------------------------------------------------------------
                 Groundwater Management Solution
------------------------------------------------------------------------


     OPTIMAL SOLUTION FOUND

     OPTIMAL RATES FOR EACH FLOW VARIABLE
     ------------------------------------------
```

Variable Name	Withdrawal Rate	Injection Rate	Contribution To Objective
Q1	0.000000E+00		0.000000E+00
Q2	2.523063E+03		0.000000E+00
Q3	1.476937E+03		0.000000E+00
Q4	0.000000E+00		0.000000E+00
Q5	0.000000E+00		0.000000E+00
TOTALS	4.000000E+03	0.000000E+00	0.000000E+00

```
     OPTIMAL VALUES FOR EACH EXTERNAL VARIABLE
```

```
          --------------------------------------
Variable              Variable           Optimal          Contribution
Name                  Type               Value            To Objective
----------            --------------     ------------     ------------
  R                   Head               4.459702E+01     4.459702E+01
                                                          ------------
TOTAL                                                     4.459702E+01
```

          OPTIMAL VALUES FOR EACH STATE VARIABLE

```
Variable                                             Contribution
Name                  Value                          To Objective
----------            ------------                   ------------
  H1                  4.459702E+01                   0.000000E+00
  H2                  5.106773E+01                   0.000000E+00
  H3                  4.459702E+01                   0.000000E+00
  H4                  4.557614E+01                   0.000000E+00
                      ------------                   ------------
TOTALS                1.858379E+02                   0.000000E+00
```

          OBJECTIVE FUNCTION VALUE   4.459702E+01

          BINDING CONSTRAINTS

```
Constraint Type          Name      Status      Shadow Price
---------------          ----      ------      ------------
Summation                DEMAND    Binding     -6.1069E-03
Summation                CON1      Binding     -5.4299E-01
Summation                CON3      Binding     -4.5701E-01
```

Binding constraint values are determined from the linear program
   and based on the response matrix approximation of the flow process.

          Range Analysis Not Reported

          Final Flow Process Simulation

  Running Final Flow Process Simulation
     using Optimal Flow Variable Rates

        Status of State Variable Values
           Using Optimal Flow Rate Variable Values

```
        State Variable Type      Name      Computed Value
        -------------------      ----      --------------
        Head                     H1        4.4597015E+01
        Head                     H2        5.1067732E+01
        Head                     H3        4.4597015E+01
        Head                     H4        4.5576141E+01
```
Precision limitations and nonlinear response may cause
   the state variables computed directly by the flow process
   to differ from those computed using the linear program.

        Status of Simulation-Based Constraints
           Using Optimal Flow Rate Variable Values

Constraint Type	Name	Simulated By Flow Process	Specified in Constraints	Difference
Summation	DEMAND	4.0000E+03 >	4.0000E+03	0.0000E+00
Summation	CON1	6.1703E-08 >	0.0000E+00	6.1703E-08
Summation	CON2	6.4707E+00 >	0.0000E+00	6.4707E+00
Summation	CON3	-8.4385E-08 >	0.0000E+00	-8.4385E-08
Summation	CON4	9.7913E-01 >	0.0000E+00	9.7913E-01

```
Difference is computed by subtracting right hand side of the constraint
  from the left side of the constraint.
Precision limitations and nonlinear response may cause the
  values of the binding constraints computed directly by the flow process
  to differ from those computed using the linear program.
```

STREAMFLOW: Maximize Summer Streamflow

The purpose of this sample problem is to demonstrate the use of streamflow-type state variables to address the problem of maximizing streamflow (or, viewed from another perspective, minimizing streamflow depletion) during periods of the year when water-supply demands are high and streamflow rates are simultaneously low.

The stream-aquifer system that is simulated is similar to that used for the SUPPLY problem described in Ahlfeld and others (2005, 2009) (fig. 5). A confined aquifer is in hydraulic connection with two streams. The aquifer is homogeneous and isotropic with a transmissivity of 5,000 ft²/d and a storage coefficient of 0.05 (dimensionless). The modeled area of interest is 6,000 ft long and 5,000 ft wide. The model consists of a single layer with 25 rows and 30 columns; each model cell is 200 ft by 200 ft. The modeled area is bounded on the east and west by no-flow conditions and on the north and south by constant heads that decrease in elevation from west to east (fig. 5). Each stream is 20 ft wide, has a streambed hydraulic conductivity of 5.0 ft/d, and has a streambed thickness of 1 ft. It is assumed that the length of each stream in each cell is 200 ft. The two streams are simulated with the MODFLOW Streamflow-Routing (SFR) Package (Prudic and others, 2004; Niswonger and Prudic, 2005).

A 3-year period of water-supply management is simulated. The 3-year period is divided into 12 seasons (winter, spring, summer, and fall of each year), each of which is represented by a single stress period. The aquifer is recharged at a rate of 0.0005 ft/d in the winter, 0.002 ft/d in the spring, 0 ft/d in the summer, and 0.001 ft/d in the fall. The variable rates of recharge result in streamflow rates for each stream that are highest during the spring and lowest during the summer.

The MODFLOW model consists of a **NAME** file, a Discretization (**DIS**) file, a Basic (**BAS**) file, a Block-Centered Flow (**BCF**) file, a Recharge (**RCH**) file, an **SFR** file, an Output Control (**OC**) file, and a Preconditioned Conjugate-Gradient (**PCG**) file.

Groundwater-Management Problem

The objective of the management problem is to maximize streamflow during the summer of the third year of simulation (stress period 11) at four locations shown in figure 5. Four streamflow-type state variables are defined for the problem ($SF1$, $SF2$, $SF3$, $SF4$) representing the streamflow at the end of the summer season, and the objective is

$$\text{Maximize } SF1 + SF2 + SF3 + SF4 \; . \tag{17}$$

Dimensionless, uniform weighting coefficients equal to 1.0 are specified for each state-variable coefficient (ξ_r), and FNTYP is set to USDV in the **OBJFNC** file, indicating that these variables are not multiplied by their duration.

Groundwater that is pumped to meet water-supply demands can be withdrawn at three candidate well locations, $Q1$, $Q2$, and $Q3$ (fig. 5). Because it may be advantageous to have some wells pump at variable rates throughout the year, multiple decision variables are defined for wells $Q2$ and $Q3$: Well $Q2$ is allowed to have a different withdrawal rate during each of the four seasons (identified as decision variables $Q2win$, $Q2spr$, $Q2sum$, and $Q2fal$), and well $Q3$ can be pumped during the spring ($Q3spr$) and summer ($Q3sum$) months. Well $Q1$ must have a constant withdrawal rate throughout the year. The minimum and maximum pumping rates at each well are 0 and 50,000 ft³/d, respectively.

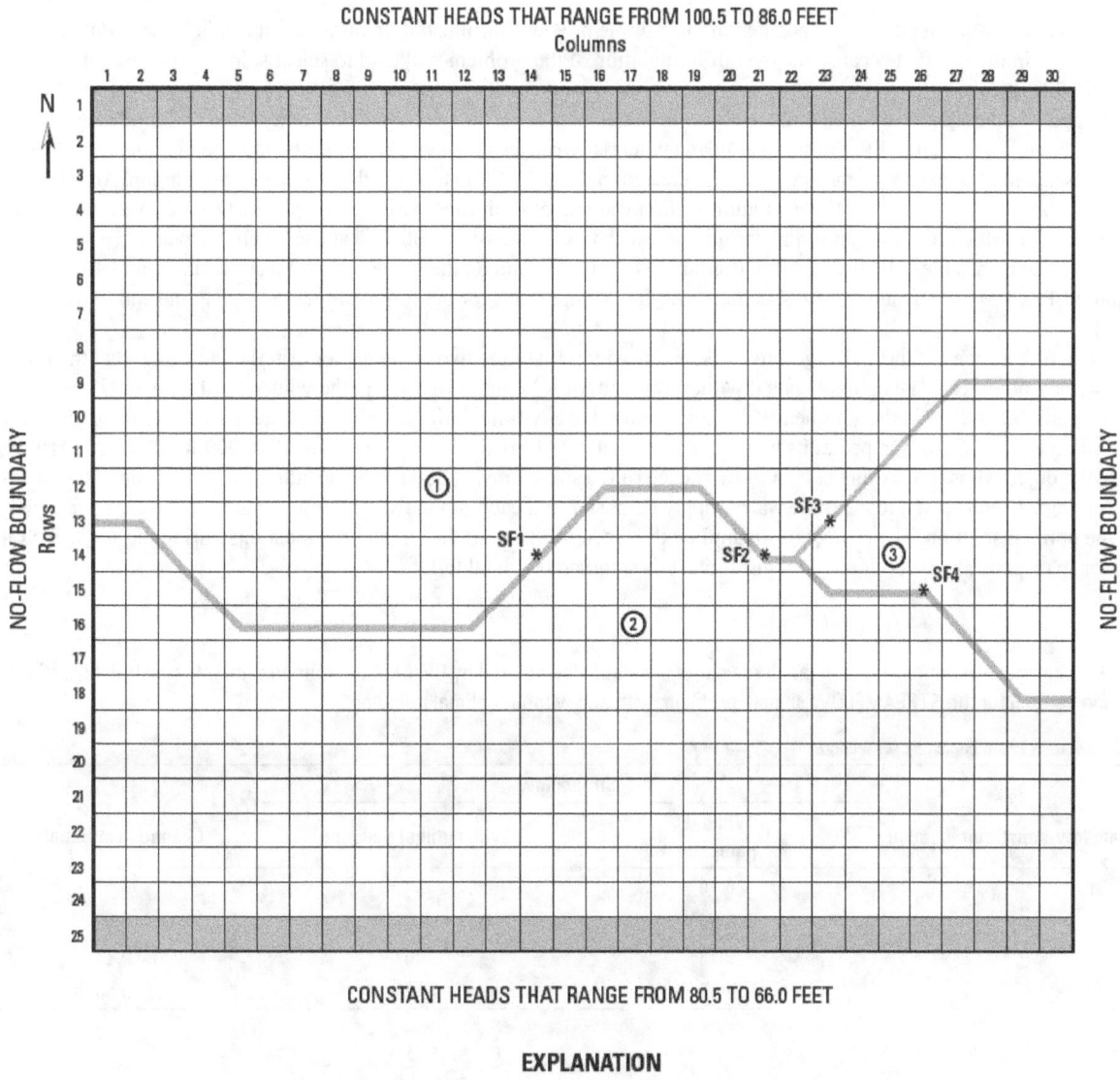

Figure 5. Schematic diagram showing model grid for STREAMFLOW sample problem.

The minimum water-supply demands on the aquifer are variable throughout the year, with rates of 30,000 ft³/d during the winter and fall, 45,000 ft³/d during the spring, and 60,000 ft³/d during the summer. The maximum demands are constant throughout the year and are equal to 80,000 ft³/d. These seasonally variable demands are specified by use of linear-summation constraints. For example, the minimum summer demand is written mathematically as

$$Q1 + Q2sum + Q3sum \geq 60,000 \cdot \tag{18}$$

Although the maximum demands are specified in this sample problem for illustrative purposes, it can be expected that they will not be binding in the solution because the overall formulation of the problem will tend to select solutions that pump at the smallest permissible rates.

Although the aquifer is confined and the objective function is linear, the presence of the two streams introduces head-dependent boundary conditions to the simulation and potentially causes the management problem to be nonlinear. As a consequence, sequential linear programming (SLP) is selected in the **SOLN** file to solve the nonlinear formulation. An initial perturbation value of 20 percent of the maximum withdrawal rate of each candidate well is specified (that is, variable DINIT is specified as 0.2), which results in an initial perturbation withdrawal rate of 10,000 ft³/d at each well. Convergence criteria on the flow-rate decision variables of 1.0×10^{-5} ft³/d (variable SLPVCRIT) and on the value of the objective function of 1.0×10^{-4} ft³ (variable SLPZCRIT) of streamflow are specified. The GWM input files for the formulation are listed at the end of the sample problem.

Only two iterations of the SLP algorithm were required to satisfy the two convergence criteria, indicating that the problem is very weakly nonlinear. The optimal annual withdrawal pattern is to pump well $Q2$ in the winter and fall, well $Q3$ in the spring, and both wells $Q2$ and $Q3$ in the summer. Well $Q1$ is not used at any time during the year. In order to meet peak demand during the summer, well $Q3$ is pumped at its maximum rate of 50,000 ft³/d and well $Q2$ at a rate of 10,000 ft³/d. Not surprisingly, because the objective is to maximize streamflow at each of the streamflow state-variable locations, the minimum amount of water is pumped each season to meet the water-supply demands, and each of the four minimum water-supply demands are binding at the optimal solution. The change in streamflow that occurs at each of the streamflow locations during the summer of the third year in response to the optimal pumping strategy is summarized in table 1.

Table 1. Summer streamflow rates in the third year of simulation calculated by GWM at the four streamflow-constraint locations shown in figure 5 for the STREAMFLOW sample problem, with and without optimal pumping.

[Streamflow rates are in cubic feet per second.]

Streamflow-constraint location	Streamflow		Change in streamflow
	Without pumping	With optimal pumping	
SF1	6.30	6.27	0.03
SF2	8.95	8.86	0.09
SF3	3.18	3.01	0.17
SF4	13.38	12.78	0.60

Selected Input and Output Files

NAME file (*streamflow.nam*)

```
LIST     10    streamflow.lst
DIS      11    ..\data\STREAMFLOW\streamflow.dis
BAS6     12    ..\data\STREAMFLOW\streamflow.ba6
BCF6     13    ..\data\STREAMFLOW\streamflow.bc6
RCH      15    ..\data\STREAMFLOW\streamflow.rch
SFR      16    ..\data\STREAMFLOW\streamflow.sfr
OC       17    ..\data\STREAMFLOW\streamflow.oc
PCG      18    ..\data\STREAMFLOW\streamflow.pcg
GWM      19    ..\data\STREAMFLOW\streamflow.gwm
```

```
DATA    20    streamflow.gwmwell
DATA    81    streamflow.sfrout
```

GWM file (*streamflow.gwm*)

```
#STREAMFLOW Sample Problem--GWM File
#December 2009
OUT      streamflow.gwmout
DECVAR   ..\data\STREAMFLOW\streamflow.decvar
STAVAR   ..\data\STREAMFLOW\streamflow.stavar
OBJFNC   ..\data\STREAMFLOW\streamflow.objfnc
VARCON   ..\data\STREAMFLOW\streamflow.varcon
SUMCON   ..\data\STREAMFLOW\streamflow.sumcon
SOLN     ..\data\STREAMFLOW\streamflow.soln
```

Decision variable (DECVAR) file (*streamflow.decvar*)

```
#STREAMFLOW Sample Problem--DECVAR File
#December 2009
 1 20                       #1-IPRN   GWMWFILE
 7 0 0                      #2-NFVAR NEVAR NBVAR
Q1    1  1  12  11  W  Y  1-12      #3a-FVNAME NC LAY ROW COL FTYPE FSTAT WSP
Q2win 1  1  16  17  W  Y  1:5:9
Q2spr 1  1  16  17  W  Y  2:6:10
Q2sum 1  1  16  17  W  Y  3:7:11
Q2fal 1  1  16  17  W  Y  4:8:12
Q3spr 1  1  14  25  W  Y  2:6:10
Q3sum 1  1  14  25  W  Y  3:7:11
```

State variables file (STAVAR) file (*streamflow.stavar*)

```
#STREAMFLOW Sample Problem--STAVAR File
#December 2009
 1                         #1-IPRN
 0  4  0                   #2-NHVAR NRVAR NSVAR
SF1  1  14  11             #4-SVNAME SEG REACH SVSP
SF2  1  21  11
SF3  2   8  11
SF4  3   5  11
```

Objective function (OBJFNC) file (*streamflow.objfnc*)

```
#STREAMFLOW Sample Problem--OBJFNC File
#December 2009
 1             #1-IPRN
MAX   USDV     #2-OBJTYP FNTYP
 0  0  0  4    #3-NFVOBJ NEVOBJ  NBVOBJ NSVOBJ
SF1 1.00       #7-SVNAME  SVOBJC
SF2 1.00
SF3 1.00
SF4 1.00
```

Decision-variable constraints (VARCON) file (*streamflow.varcon*)

```
#STREAMFLOW Sample Problem--VARCON File
#December 2009
```

```
1                                #1-IPRN
Q1    0.0   5.0d4   0.0D2        #2-FVNAME FVMIN FVMAX FVREF
Q2win 0.0   5.0d4   0.0D2
Q2spr 0.0   5.0d4   0.0D2
Q2sum 0.0   5.0d4   0.0D2
Q2fal 0.0   5.0d4   0.0D2
Q3spr 0.0   5.0d4   0.0D2
Q3sum 0.0   5.0d4   0.0D2
```

Linear-summation constraints (SUMCON) file (*streamflow.sumcon*)

```
#STREAMFMINW Sample Problem--SUMCON File
#December 2009
 1                              #1-IPRN
 8                              #2-SMCNUM
WIN-MAX 2 le 80000.     #3a-SMCNAME NTERMS TYPE RHS
 Q1   1.0               #3b-GVNAME GVCOEFF
 Q2win 1.0
WIN-MIN 2 ge 30000.
 Q1   1.0
 Q2win 1.0
SPR-MAX 3 le 80000.
 Q1   1.0
 Q2spr 1.0
 Q3spr 1.0
SPR-MIN  3 ge 45000.
 Q1   1.0
 Q2spr 1.0
 Q3spr 1.0
SUM-MAX  3 le 80000.
 Q1   1.0
 Q2sum 1.0
 Q3sum 1.0
SUM-MIN  3 ge 60000.
 Q1   1.0
 Q2sum 1.0
 Q3sum 1.0
FAL-MAX  2 le 80000.
 Q1   1.0
 Q2fal 1.0
FAL-MIN  2 ge 30000.
 Q1   1.0
 Q2fal 1.0
```

Solution and output control (SOLN) file (*streamflow.soln*)

```
#STREAMFLOW Sample Problem--SOLN File
#December 2009
 SLP                            #1-SOLNTYP
 50 10000    2000               #5a-SLPITMAX LPITMAX BBITMAX
0.00001  0.0001  0.2  0.00002 2 #5b-SLPVCRIT SLPZCRIT DINIT DMIN DSC
1   10  0.5  0.5    5    0.0    #5c-NSIGDIG NPGNMX PGFACT AFACT NINFMX CRITMFC
 1 1 0                          #5d-SLPITPRT BBITPRT RANGE
 0                              #6a-IBASE
```

Part of the GWM-2005 output file (*streamflow.gwmout*)

```
-----------------------------------------------------------------------------
                 Groundwater Management Solution
-----------------------------------------------------------------------------


      OPTIMAL SOLUTION FOUND

      OPTIMAL RATES FOR EACH FLOW VARIABLE
      -------------------------------------------
Variable           Withdrawal           Injection          Contribution
Name               Rate                 Rate               To Objective
----------         --------------       ------------       ------------
  Q1               0.000000E+00                            0.000000E+00
  Q2win            3.000000E+04                            0.000000E+00
  Q2spr            0.000000E+00                            0.000000E+00
  Q2sum            1.000000E+04                            0.000000E+00
  Q2fal            3.000000E+04                            0.000000E+00
  Q3spr            4.500000E+04                            0.000000E+00
  Q3sum            5.000000E+04                            0.000000E+00
                   ------------         ------------       ------------
TOTALS             1.650000E+05         0.000000E+00       0.000000E+00

      OPTIMAL VALUES FOR EACH STATE VARIABLE
      -------------------------------------------
Variable                                                   Contribution
Name               Value                                   To Objective
----------         ------------                            ------------
  SF1              5.416810E+05                            5.416810E+05
  SF2              7.655867E+05                            7.655867E+05
  SF3              2.597073E+05                            2.597073E+05
  SF4              1.104444E+06                            1.104444E+06
                   ------------                            ------------
TOTALS             2.671419E+06                            2.671419E+06


      OBJECTIVE FUNCTION VALUE                             2.671419E+06

      BINDING CONSTRAINTS
Constraint Type          Name      Status      Shadow Price
---------------          ----      ------      ------------
Summation                WIN-MIN   Binding     -3.0932E-03
Summation                SPR-MIN   Binding     -1.7106E-02
Summation                SUM-MIN   Binding     -1.6057E+00
Summation                FAL-MIN   Binding     -1.5099E-04
Maximum Flow Rate        Q3sum     Binding     Not Available

   Binding constraint values are determined from the linear program
     and based on the response matrix approximation of the flow process.

      Range Analysis Not Reported
-----------------------------------------------------------------------------
      Final Flow Process Simulation
-----------------------------------------------------------------------------
   Running Final Flow Process Simulation
     using Optimal Flow Variable Rates
```

```
Status of State Variable Values
   Using Optimal Flow Rate Variable Values
   State Variable Type     Name        Computed Value
   -------------------     ----        --------------
   Streamflow              SF1         5.4168096E+05
   Streamflow              SF2         7.6558675E+05
   Streamflow              SF3         2.5970728E+05
   Streamflow              SF4         1.1044439E+06
Precision limitations and nonlinear response may cause
   the state variables computed directly by the flow process
   to differ from those computed using the linear program.


   Status of Simulation-Based Constraints
      Using Optimal Flow Rate Variable Values


                                       Simulated    Specified
                                       By Flow         in
   Constraint Type         Name        Process      Constraints     Difference
   ---------------         ----        ----------   -----------     ----------
   Summation               WIN-MAX     3.0000E+04 < 8.0000E+04     -5.0000E+04
   Summation               WIN-MIN     3.0000E+04 > 3.0000E+04      0.0000E+00
   Summation               SPR-MAX     4.5000E+04 < 8.0000E+04     -3.5000E+04
   Summation               SPR-MIN     4.5000E+04 > 4.5000E+04      0.0000E+00
   Summation               SUM-MAX     6.0000E+04 < 8.0000E+04     -2.0000E+04
   Summation               SUM-MIN     6.0000E+04 > 6.0000E+04      0.0000E+00
   Summation               FAL-MAX     3.0000E+04 < 8.0000E+04     -5.0000E+04
   Summation               FAL-MIN     3.0000E+04 > 3.0000E+04      0.0000E+00

Difference is computed by subtracting right hand side of the constraint
   from the left side of the constraint.
Precision limitations and nonlinear response may cause the
   values of the binding constraints computed directly by the flow process
   to differ from those computed using the linear program.
```

STORAGE: Control Changes in Aquifer Storage

The STORAGE sample problem is designed to demonstrate the use of storage state variables in the STA Package. The management-problem formulation seeks to maximize withdrawals while controlling the rate of storage depletion. The hypothetical aquifer system is based on the sample problem described in appendix D of McDonald and Harbaugh (1988). The aquifer system consists of an upper unconfined aquifer and two underlying confined aquifers. The aquifers are separated by confining layers. The model consists of three layers to represent the three aquifers; each layer contains 15 rows and 15 columns, and each cell measures 5,000 ft on each side (fig. 6). Flow within the confining layers is not simulated, but the effects of the confining layers on flow between the active layers are incorporated in the vertical leakance terms between layers. Flow into the system consists of recharge from precipitation; flow out of the system is to buried drains in layer 1, discharging wells, and a lake that is represented by constant-head boundary conditions along the western side of the model in layers 1 and 2 (fig. 6). The original steady-state flow system described in McDonald and Harbaugh (1988) was converted to a transient system for the STORAGE sample problem. The model consists of six stress periods, an initial steady-state stress period representing premanagement conditions followed by five 10-year transient stress periods. The model uses the Layer Property Flow (LPF) Package rather than the BCF6 Package. The LPF input is constructed to mimic the behavior of the original problem, with very large vertical hydraulic conductivity specified for each of the three aquifers so that the vertical hydraulic conductivity in the confining layers dominates flow between the layers. The units of the model were changed from foot-second to foot-days. Layer 1 uses a specific yield of 0.15, whereas layers 2 and 3 are assigned a specific storage of 1.0×10^{-5} ft^{-1}. Both the recharge and premanagement pumping rates specified in the original problem were reduced by a factor of 10 to represent an arid climate and increase the importance of storage for the problem.

The MODFLOW model consists of a **NAME** file, a Discretization (**DIS**) file, a Basic (**BAS**) file, an LPF file, a Recharge (**RCH**) file, a Well (**WEL**) file, a Drain (**DRN**) file, an Output Control (**OC**) file, and a Strongly Implicit Procedure (**SIP**) file.

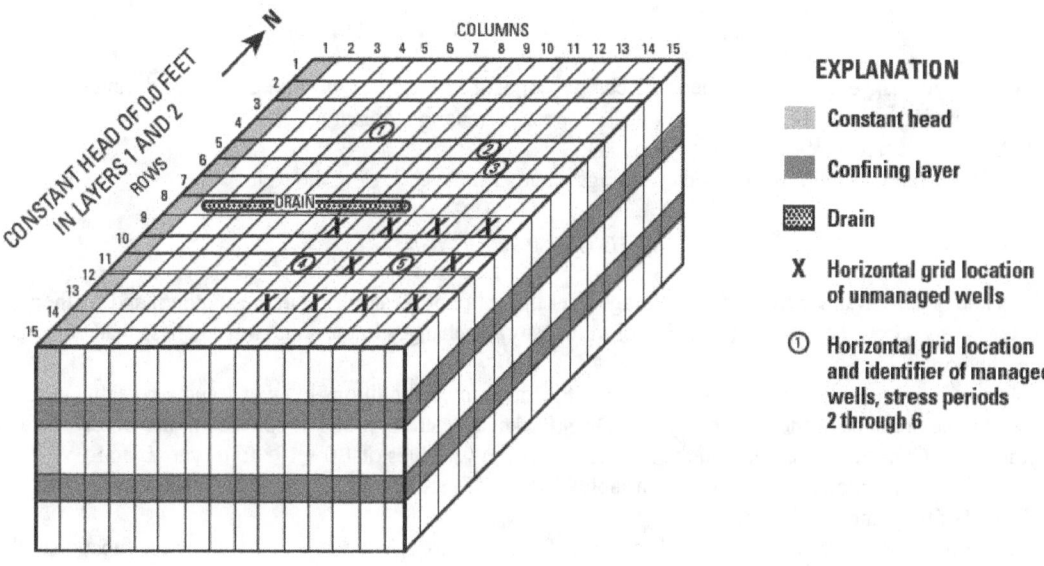

Figure 6. Schematic diagram showing model grid for STORAGE sample problem. (Modified from McDonald and Harbaugh, 1988.)

Groundwater-Management Problem

A total of 15 wells pump during the premanagement, steady-state stress period. These consist of 3 wells that pump from the confined aquifers in the northern part of the model domain and 12 wells that pump from a well field in the unconfined aquifer in the sourthern part of the domain (fig. 6). The three wells in the north (wells 1–3, fig. 6) are converted to decision variables for the 50-year management period. Ten of the 12 southern wells continue to pump during the 50-year management period at rates that are specified in the **WEL** input file; that is, the 10 wells are unmanaged and their pumping rates are not determined by GWM. Decision variables are defined to determine optimal pumping rates at the remaining two well sites (well sites 4 and 5, fig. 6), with the added possibility of pumping from the confined aquifers (layers 2 and 3) at these same horizontal locations. Therefore, there are a total of six candidate well locations in the southern part of the model—three at well site 4 and three at well site 5. Because pumping rates are allowed to vary at each of the nine candidate pumping locations during the five transient stress periods, there are a total of 45 decision variables, as shown in the **DECVAR** file at the end of the sample problem. The maximum pumping rate at each well is specified in the **VARCON** input file at about 1.5 times the steady-state pumping rate at each well.

The objective function for the management problem is to maximize the total water withdrawn at the nine withdrawal locations during the first management stress period (that is, the second GWF Process stress period) (see **OBJFNC** input file at end of sample problem). At the same time, summation constraints are imposed that require the total pumping rate in each of the five management stress periods to be the same. In effect, the objective, along with these constraints, finds the maximum sustained withdrawal that can be achieved over the 50-year simulation period.

The optimal withdrawal rates are constrained by limits on changes in aquifer storage. To create these storage constraints, five state variables are defined that represent the change in storage in the top layer of the model over each 10-year stress period. An additional two state variables define the storage change over the entire 50-year planning horizon for the north and south portions of layer 2. These state variables are defined in the **STAVAR** input file; zone information is read from both the **STAVAR** input file and from the external file *storage.zone1*. The example **STAVAR** file demonstrates the different ways in which storage-zone information can be specified in GWM.

The storage state variables are used in summation constraints (see **SUMCON** input file) that are imposed to require that the reduction in storage in each stress period be a fraction of the change in storage in the previous stress period. For example, using variable *SCHNG2* to define the change in storage in layer 1 from the beginning to the end of stress period 2 and *SCHNG3* to define the change in storage in layer 1 during stress period 3, the constraint

$$SCHNG3 \leq 0.85SCHNG2 \tag{19}$$

requires that the storage change during stress period 3 can be at most 85 percent of the storage change during stress period 2. Similar constraints are imposed on each subsequent pair of stress periods so that storage changes are required to decrease by no more than 80, 75, and 70 percent, respectively.

An additional, similar constraint is imposed on storage changes in layer 2 using the form

$$SCHNGS \leq 0.90SCHNGN \,, \tag{20}$$

where *SCHNGS* and *SCHNGN* are the changes in storage over the entire 50-year period in the southern and northern portions of the layer 2, respectively. Here, the requirement is imposed that storage changes in the south must be less than 90 percent of storage changes in the north.

The problem contains a nonlinear response because of the unconfined conditions in layer 1 and a potential nonlinear response caused by the presence of the head-dependent DRN Package. Therefore, SLP is selected in the **SOLN** file to solve the nonlinear formulation. Convergence criteria on the flow-rate decision variables of 1.0×10^{-5} ft^3/d (variable SLPVCRIT) and on the value of the objective function of 1.0×10^{-4} ft^3 (variable SLPZCRIT) are specified. The GWM input files for the formulation are listed at the end of the sample problem.

The SLP algorithm converges with an optimal solution in four iterations. The value of the objective function at the solution is 1.258205×10^9 ft^3 of water withdrawn during the first management period (stress period 2), which is equivalent to 344,714 ft^3/d. Water is withdrawn at this same rate for the entire 50-year management horizon, as indicated in the solution-report section of the GWM **OUT** file and in the volumetric-budget sections of the MODFLOW **LIST** file for stress periods 2 through 6 (see the MANAGED FLOW row in the outflow section of each volumetric budget). Although the total amount of water withdrawn each stress period is the same, the set of active pumping wells varies from one stress period to the next. GWM chooses to pump all three northern wells at their maximum rates in all stress periods; pumping rates at the southern wells, however, vary among the wells, with most of the withdrawals occurring from the western three wells. All storage-change constraints are binding at the solution except the first one, which is the storage-change requirement in layer 1 of the model between the first and second management stress periods (stress periods 2 and 3 of the model).

Selected Input and Output Files

NAME file (*storage.nam*)

```
# STORAGE sample problem for GWM
LIST  7 storage.lst
BAS6  8 ..\data\STORAGE\storage.ba6
LPF  11 ..\data\STORAGE\storage.lpf
WEL  12 ..\data\STORAGE\storage.wel
DRN  13 ..\data\STORAGE\storage.drn
RCH  18 ..\data\STORAGE\storage.rch
SIP  19 ..\data\STORAGE\storage.sip
OC   22 ..\data\STORAGE\storage.oc
DIS  10 ..\data\STORAGE\storage.dis
GWM  20 ..\data\STORAGE\storage.gwm
DATA 75 storage.gwmwell
```

GWM file (*storage.gwm*)

```
# STORAGE sample problem
OUT     storage.gwmout
DECVAR  ..\data\STORAGE\storage.decvar
STAVAR  ..\data\STORAGE\storage.stavar
OBJFNC  ..\data\STORAGE\storage.objfnc
VARCON  ..\data\STORAGE\storage.varcon
```

```
SUMCON   ..\data\STORAGE\storage.sumcon
SOLN     ..\data\STORAGE\storage.soln
```

Decision variable (DECVAR) file (*storage.decvar*)

```
#STORAGE Sample Problem, DECVAR file
 1 75                              #1-IPRN  GWMWFILE
 45 0 0                            #2-NFVAR  NEVAR  NBVAR
Q1-2-a    1    2    4     6    W   Y   2 #FVNAME NC LAY ROW COL FTYPE FSTAT WSP
Q1-2-b    1    2    4     6    W   Y   3 # Naming convention
Q1-2-c    1    2    4     6    W   Y   4 # Q_ refers to well number
Q1-2-d    1    2    4     6    W   Y   5 # -_- refers to layer of well
Q1-2-e    1    2    4     6    W   Y   6 # suffix a = stress period 2
Q2-3-a    1    3    5    11    W   Y   2 # suffix b = stress period 3
Q2-3-b    1    3    5    11    W   Y   3 # suffix c = stress period 4
Q2-3-c    1    3    5    11    W   Y   4 # suffix d = stress period 5
Q2-3-d    1    3    5    11    W   Y   5 # suffix e = stress period 6
Q2-3-e    1    3    5    11    W   Y   6
Q3-2-a    1    2    6    12    W   Y   2
Q3-2-b    1    2    6    12    W   Y   3
Q3-2-c    1    2    6    12    W   Y   4
Q3-2-d    1    2    6    12    W   Y   5
Q3-2-e    1    2    6    12    W   Y   6
Q4-1-a    1    1   11     8    W   Y   2
Q4-1-b    1    1   11     8    W   Y   3
Q4-1-c    1    1   11     8    W   Y   4
Q4-1-d    1    1   11     8    W   Y   5
Q4-1-e    1    1   11     8    W   Y   6
Q4-2-a    1    2   11     8    W   Y   2
Q4-2-b    1    2   11     8    W   Y   3
Q4-2-c    1    2   11     8    W   Y   4
Q4-2-d    1    2   11     8    W   Y   5
Q4-2-e    1    2   11     8    W   Y   6
Q4-3-a    1    3   11     8    W   Y   2
Q4-3-b    1    3   11     8    W   Y   3
Q4-3-c    1    3   11     8    W   Y   4
Q4-3-d    1    3   11     8    W   Y   5
Q4-3-e    1    3   11     8    W   Y   6
Q5-1-a    1    1   11    12    W   Y   2
Q5-1-b    1    1   11    12    W   Y   3
Q5-1-c    1    1   11    12    W   Y   4
Q5-1-d    1    1   11    12    W   Y   5
Q5-1-e    1    1   11    12    W   Y   6
Q5-2-a    1    2   11    12    W   Y   2
Q5-2-b    1    2   11    12    W   Y   3
Q5-2-c    1    2   11    12    W   Y   4
Q5-2-d    1    2   11    12    W   Y   5
Q5-2-e    1    2   11    12    W   Y   6
Q5-3-a    1    3   11    12    W   Y   2
Q5-3-b    1    3   11    12    W   Y   3
Q5-3-c    1    3   11    12    W   Y   4
Q5-3-d    1    3   11    12    W   Y   5
Q5-3-e    1    3   11    12    W   Y   6
```

State variables file (STAVAR) file (*storage.stavar*)

```
#STORAGE Sample Problem, STAVAR file
1                          #1-IPRN
0   0   7                  #2-NHVAR NRVAR NSVAR
SCHNG2  2  2  zone         # SVNAME SPSTRT SPEND CZONE:  Layer 1, stress period 2
1                          # NSVZL: 1 layer in zone
1                          # LNUM : 1st layer is layer 1
CONSTANT 1                 # SVZONE: zone is all of layer 1
SCHNG3  3  3  zone         # SVNAME SPSTRT SPEND CZONE:  Layer 1, stress period 3
1                          # NSVZL
1                          # LNUM
CONSTANT 1                 # SVZONE - zone is all of layer 1
SCHNG4  4  4  zone         # SVNAME SPSTRT SPEND CZONE:  Layer 1, stress period 4
1                          # NSVZL
1                          # LNUM
CONSTANT 1                 # SVZONE - zone is all of layer 1
SCHNG5  5  5  zone         # SVNAME SPSTRT SPEND CZONE:  Layer 1, stress period 5
1                          # NSVZL
1                          # LNUM
CONSTANT 1                 # SVZONE - zone is all of layer 1
SCHNG6  6  6  zone         # SVNAME SPSTRT SPEND CZONE:  Layer 1, stress period 6
1                          # NSVZL
1                          # LNUM
CONSTANT 1                 # SVZONE - zone is all of layer 1
SCHNGN  2  6  zone         # SVNAME SPSTRT SPEND CZONE:  Layer 1, stress period 6
1                          # NSVZL
2                          # LNUM
OPEN/CLOSE ..\data\STORAGE\storage.zone1  1  (15I4)   8 # SVZONE-north part,layer 2
SCHNGS  2  6  zone         # SVNAME SPSTRT SPEND CZONE:  stress period 2 to 6
1                          # NSVZL: 1 layer in zone
2                          # LNUM : 1st layer is layer 2
INTERNAL    1  (15I4)   8  # SVZONE - zone is south part,layer 2
  0  0  0  0  0  0  0  0  0  0  0  0  0  0  0
  0  0  0  0  0  0  0  0  0  0  0  0  0  0  0
  0  0  0  0  0  0  0  0  0  0  0  0  0  0  0
  0  0  0  0  0  0  0  0  0  0  0  0  0  0  0
  0  0  0  0  0  0  0  0  0  0  0  0  0  0  0
  0  0  0  0  0  0  0  0  0  0  0  0  0  0  0
  0  0  0  0  0  0  0  0  0  0  0  0  0  0  0
  0  0  0  0  0  0  0  0  0  0  0  0  0  0  0
  1  1  1  1  1  1  1  1  1  1  1  1  1  1  1
  1  1  1  1  1  1  1  1  1  1  1  1  1  1  1
  1  1  1  1  1  1  1  1  1  1  1  1  1  1  1
  1  1  1  1  1  1  1  1  1  1  1  1  1  1  1
  1  1  1  1  1  1  1  1  1  1  1  1  1  1  1
  1  1  1  1  1  1  1  1  1  1  1  1  1  1  1
  1  1  1  1  1  1  1  1  1  1  1  1  1  1  1
```

Objective function (OBJFNC) file (*storage.objfnc*)

```
#STORAGE Sample Problem, OBJFNC file
 1              #1-IPRN
 MAX   WSDV     #2-OBJTYP  FNTYP
 9  0  0  0     #3-NFVOBJ  NEVOBJ  NBVOBJ  (NSVOBJ)
Q1-2-a 1.0      #4-FVNAME  FVOBJC
```

```
Q2-3-a 1.0
Q3-2-a 1.0
Q4-1-a 1.0
Q4-2-a 1.0
Q4-3-a 1.0
Q5-1-a 1.0
Q5-2-a 1.0
Q5-3-a 1.0
```

Decision-variable constraints (VARCON) file (*storage.varcon*)

```
#STORAGE Sample Problem, VARCON file
 1                        #1-IPRN
Q1-2-a  0.0D0   6.4D4   0.0D0  #2-FVNAME  FVMIN  FVMAX  FVREF
Q1-2-b  0.0D0   6.4D4   0.0D0
Q1-2-c  0.0D0   6.4D4   0.0D0
Q1-2-d  0.0D0   6.4D4   0.0D0
Q1-2-e  0.0D0   6.4D4   0.0D0


35 input lines deleted here


Q5-3-a  0.0D0   6.4D4   0.0D0
Q5-3-b  0.0D0   6.4D4   0.0D0
Q5-3-c  0.0D0   6.4D4   0.0D0
Q5-3-d  0.0D0   6.4D4   0.0D0
Q5-3-e  0.0D0   6.4D4   0.0D0
```

Linear-summation constraints (SUMCON) file (*storage.sumcon*)

```
#STORAGE Sample Problem, SUMCON file
 1                        #1-IPRN
 9                        #2-SMCNUM
 ST1-2  2 le 0.           #3a-SMCNAME NTERMS TYPE RHS
  SCHNG2 -0.85            #3b-GVNAME GVCOEFF
  SCHNG3  1.0
 ST1-3  2 le 0.           #3a-SMCNAME NTERMS TYPE RHS
  SCHNG3 -0.80            #3b-GVNAME GVCOEFF
  SCHNG4  1.0
 ST1-4  2 le 0.           #3a-SMCNAME NTERMS TYPE RHS
  SCHNG4 -0.75            #3b-GVNAME GVCOEFF
  SCHNG5  1.0
 ST1-5  2 le 0.           #3a-SMCNAME NTERMS TYPE RHS
  SCHNG5 -0.70            #3b-GVNAME GVCOEFF
  SCHNG6  1.0
 ST2-1  2 le 0.           #3a-SMCNAME NTERMS TYPE RHS
  SCHNGS  1.0             #3b-GVNAME GVCOEFF
  SCHNGN -0.9
 EQ-b  18 eq 0.           #The next 4 constraints impose the requirement
 Q1-2-a    1.0            #that total withdrawal rates are equal during
 Q2-3-a    1.0            #each of the five 10-year management stress
 Q3-2-a    1.0            #periods. This first constraint requires total
 Q4-1-a    1.0            #withdrawals in the second management stress
 Q4-2-a    1.0            #period to be equal to those in the first.
 Q4-3-a    1.0
 Q5-1-a    1.0
 Q5-2-a    1.0
```

```
Q5-3-a     1.0
Q1-2-b    -1.0
Q2-3-b    -1.0
Q3-2-b    -1.0
Q4-1-b    -1.0
Q4-2-b    -1.0
Q4-3-b    -1.0
Q5-1-b    -1.0
Q5-2-b    -1.0
Q5-3-b    -1.0
 EQ-c   18 eq 0.
Q1-2-a     1.0
Q2-3-a     1.0
Q3-2-a     1.0
Q4-1-a     1.0
Q4-2-a     1.0
Q4-3-a     1.0
Q5-1-a     1.0
Q5-2-a     1.0
Q5-3-a     1.0
Q1-2-c    -1.0
Q2-3-c    -1.0
Q3-2-c    -1.0
Q4-1-c    -1.0
Q4-2-c    -1.0
Q4-3-c    -1.0
Q5-1-c    -1.0
Q5-2-c    -1.0
Q5-3-c    -1.0
 EQ-d   18 eq 0.
Q1-2-a     1.0
Q2-3-a     1.0
Q3-2-a     1.0
Q4-1-a     1.0
Q4-2-a     1.0
Q4-3-a     1.0
Q5-1-a     1.0
Q5-2-a     1.0
Q5-3-a     1.0
Q1-2-d    -1.0
Q2-3-d    -1.0
Q3-2-d    -1.0
Q4-1-d    -1.0
Q4-2-d    -1.0
Q4-3-d    -1.0
Q5-1-d    -1.0
Q5-2-d    -1.0
Q5-3-d    -1.0
 EQ-e   18 eq 0.
Q1-2-a     1.0
Q2-3-a     1.0
Q3-2-a     1.0
Q4-1-a     1.0
Q4-2-a     1.0
Q4-3-a     1.0
Q5-1-a     1.0
```

```
Q5-2-a    1.0
Q5-3-a    1.0
Q1-2-e   -1.0
Q2-3-e   -1.0
Q3-2-e   -1.0
Q4-1-e   -1.0
Q4-2-e   -1.0
Q4-3-e   -1.0
Q5-1-e   -1.0
Q5-2-e   -1.0
Q5-3-e   -1.0
```

Solution and output control (SOLN) file (*storage.soln*)

```
#STORAGE Sample Problem, SOLN file
  SLP                            #1-SOLNTYP
 50 10000    2000                #5a-SLPITMAX LPITMAX BBITMAX
0.00001  0.0001  0.2  0.00002 2  #5b-SLPVCRIT SLPZCRIT DINIT DMIN DSC
 1   10   0.5   0.5    5         #5c-NSIGDIG NPGNMX PGFACT AFACT NINFMX
  2  1 0                         #5d-SLPITPRT BBITPRT RANGE
  0                              #6a-IBASE
```

Part of the GWM-2005 output file (*storage.out*)

```
---------------------------------------------------------------------------
          Groundwater Management Solution
---------------------------------------------------------------------------


      OPTIMAL SOLUTION FOUND

      OPTIMAL RATES FOR EACH FLOW VARIABLE
      -------------------------------------------
```

Variable Name	Withdrawal Rate	Injection Rate	Contribution To Objective
Q1-2-a	6.400000E+04		2.336000E+08
Q1-2-b	6.400000E+04		0.000000E+00
Q1-2-c	6.400000E+04		0.000000E+00
Q1-2-d	6.400000E+04		0.000000E+00
Q1-2-e	6.400000E+04		0.000000E+00
Q2-3-a	6.400000E+04		2.336000E+08
Q2-3-b	6.400000E+04		0.000000E+00
Q2-3-c	6.400000E+04		0.000000E+00
Q2-3-d	6.400000E+04		0.000000E+00
Q2-3-e	6.400000E+04		0.000000E+00
Q3-2-a	6.400000E+04		2.336000E+08
Q3-2-b	6.400000E+04		0.000000E+00
Q3-2-c	6.400000E+04		0.000000E+00
Q3-2-d	6.400000E+04		0.000000E+00
Q3-2-e	6.400000E+04		0.000000E+00
Q4-1-a	0.000000E+00		0.000000E+00
Q4-1-b	0.000000E+00		0.000000E+00
Q4-1-c	0.000000E+00		0.000000E+00
Q4-1-d	2.471372E+04		0.000000E+00
Q4-1-e	6.400000E+04		0.000000E+00
Q4-2-a	2.119508E+04		7.736204E+07

```
Q4-2-b          3.287872E+04                              0.000000E+00
Q4-2-c          3.171860E+04                              0.000000E+00
Q4-2-d          6.400000E+04                              0.000000E+00
Q4-2-e          2.471372E+04                              0.000000E+00
Q4-3-a          6.400000E+04                              2.336000E+08
Q4-3-b          6.400000E+04                              0.000000E+00
Q4-3-c          6.400000E+04                              0.000000E+00
Q4-3-d          6.400000E+04                              0.000000E+00
Q4-3-e          6.400000E+04                              0.000000E+00
Q5-1-a          0.000000E+00                              0.000000E+00
Q5-1-b          0.000000E+00                              0.000000E+00
Q5-1-c          0.000000E+00                              0.000000E+00
Q5-1-d          0.000000E+00                              0.000000E+00
Q5-1-e          0.000000E+00                              0.000000E+00
Q5-2-a          3.518642E+03                              1.284304E+07
Q5-2-b          0.000000E+00                              0.000000E+00
Q5-2-c          0.000000E+00                              0.000000E+00
Q5-2-d          0.000000E+00                              0.000000E+00
Q5-2-e          0.000000E+00                              0.000000E+00
Q5-3-a          6.400000E+04                              2.336000E+08
Q5-3-b          5.583500E+04                              0.000000E+00
Q5-3-c          5.699512E+04                              0.000000E+00
Q5-3-d          0.000000E+00                              0.000000E+00
Q5-3-e          0.000000E+00                              0.000000E+00
              ------------    ------------              ------------
TOTALS          1.723569E+06    0.000000E+00              1.258205E+09


            OPTIMAL VALUES FOR EACH STATE VARIABLE
            ---------------------------------------

Variable                                              Contribution
Name            Value                                 To Objective
----------      ------------                          ------------

 SCHNG2         4.181905E+08                           0.000000E+00
 SCHNG3         3.320629E+08                           0.000000E+00
 SCHNG4         2.656503E+08                           0.000000E+00
 SCHNG5         1.992377E+08                           0.000000E+00
 SCHNG6         1.394664E+08                           0.000000E+00
 SCHNGN         4.795312E+06                           0.000000E+00
 SCHNGS         4.315780E+06                           0.000000E+00
              ------------                            ------------
TOTALS          1.363719E+09                           0.000000E+00


            OBJECTIVE FUNCTION VALUE                   1.258205E+09


        BINDING CONSTRAINTS
Constraint Type         Name    Status      Shadow Price
---------------         ----    ------      ------------
Summation               ST1-3   Binding      1.6353E-02
Summation               ST1-4   Binding      4.4970E-02
Summation               ST1-5   Binding      3.9403E-01
Summation               ST2-1   Binding      1.7836E+02
Summation               EQ-b    Binding      8.6442E+01
Summation               EQ-c    Binding      2.5799E+02
Summation               EQ-d    Binding     -2.7076E+02
Summation               EQ-e    Binding      3.5251E+03
```

Maximum Flow Rate	Q1-2-a	Binding	Not Available
Maximum Flow Rate	Q1-2-b	Binding	Not Available
Maximum Flow Rate	Q1-2-c	Binding	Not Available
Maximum Flow Rate	Q1-2-d	Binding	Not Available
Maximum Flow Rate	Q1-2-e	Binding	Not Available
Maximum Flow Rate	Q2-3-a	Binding	Not Available
Maximum Flow Rate	Q2-3-b	Binding	Not Available
Maximum Flow Rate	Q2-3-c	Binding	Not Available
Maximum Flow Rate	Q2-3-d	Binding	Not Available
Maximum Flow Rate	Q2-3-e	Binding	Not Available
Maximum Flow Rate	Q3-2-a	Binding	Not Available
Maximum Flow Rate	Q3-2-b	Binding	Not Available
Maximum Flow Rate	Q3-2-c	Binding	Not Available
Maximum Flow Rate	Q3-2-d	Binding	Not Available
Maximum Flow Rate	Q3-2-e	Binding	Not Available
Maximum Flow Rate	Q4-1-e	Binding	Not Available
Maximum Flow Rate	Q4-2-d	Binding	Not Available
Maximum Flow Rate	Q4-3-a	Binding	Not Available
Maximum Flow Rate	Q4-3-b	Binding	Not Available
Maximum Flow Rate	Q4-3-c	Binding	Not Available
Maximum Flow Rate	Q4-3-d	Binding	Not Available
Maximum Flow Rate	Q4-3-e	Binding	Not Available
Maximum Flow Rate	Q5-3-a	Binding	Not Available

Binding constraint values are determined from the linear program
and based on the response matrix approximation of the flow process.

        Range Analysis Not Reported

        Final Flow Process Simulation

Running Final Flow Process Simulation
  using Optimal Flow Variable Rates

    Status of State Variable Values
      Using Optimal Flow Rate Variable Values

State Variable Type	Name	Computed Value
Change in Storage	SCHNG2	4.1819050E+08
Change in Storage	SCHNG3	3.3206289E+08
Change in Storage	SCHNG4	2.6565031E+08
Change in Storage	SCHNG5	1.9923773E+08
Change in Storage	SCHNG6	1.3946641E+08
Change in Storage	SCHNGN	4.7953115E+06
Change in Storage	SCHNGS	4.3157804E+06

Precision limitations and nonlinear response may cause
  the state variables computed directly by the flow process
  to differ from those computed using the linear program.

    Status of Simulation-Based Constraints
      Using Optimal Flow Rate Variable Values

		Simulated By Flow	Specified in	
Constraint Type	Name	Process	Constraints	Difference

```
    ----------------    ----        ----------    ----------    ----------
    Summation           ST1-2       -2.3399E+07 < 0.0000E+00   -2.3399E+07
    Summation           ST1-3       -6.8939E+00 < 0.0000E+00   -6.8939E+00
    Summation           ST1-4        5.6488E+00 < 0.0000E+00    5.6488E+00
    Summation           ST1-5        4.5553E-01 < 0.0000E+00    4.5553E-01
    Summation           ST2-1        6.4611E-03 < 0.0000E+00    6.4611E-03
    Summation           EQ-b         2.1828E-11 = 0.0000E+00    2.1828E-11
    Summation           EQ-c         2.1828E-11 = 0.0000E+00    2.1828E-11
    Summation           EQ-d         2.9104E-11 = 0.0000E+00    2.9104E-11
    Summation           EQ-e         2.1828E-11 = 0.0000E+00    2.1828E-11
```

Difference is computed by subtracting right hand side of the constraint
   from the left side of the constraint.
Precision limitations and nonlinear response may cause the
   values of the binding constraints computed directly by the flow process
   to differ from those computed using the linear program.

Acknowledgments

The authors thank Mikaela Martin Laverty, University of Massachusetts at Amherst, for her extensive testing of the STA Package. The authors also thank Jennifer Stanton, Brian Wagner, and Kim Otto of the U.S. Geological Survey for their helpful review comments on a draft of this report.

References Cited

Ahlfeld, D.P., Baker, K.M., and Barlow, P.M., 2009, GWM-2005—A Groundwater-Management Process for MODFLOW-2005 with Local Grid Refinement (LGR) capability: U.S. Geological Survey Techniques and Methods 6-A33, 65 p.

Ahlfeld, D.P., Barlow, P.M., and Mulligan, A.E., 2005, GWM—A Ground-Water Management Process for the U.S. Geological Survey modular ground-water model (MODFLOW-2000): U.S. Geological Survey Open-File Report 2005–1072, 124 p.

Ahlfeld, D.P., and Mulligan, A.E., 2000, Optimal management of flow in groundwater systems: San Diego, CA, Academic Press, 185 p.

Anderman, E.R., and Hill, M.C., 2000, MODFLOW-2000, the U.S. Geological Survey modular ground-water model—documentation of the Hydrogeologic-Unit Flow (HUF) Package: U.S. Geological Survey Open-File Report 00-342, 89 p.

Baker, Kristine, 2008, Extension of MF2005-GWM (Ground-Water Management) to solve management formulations which optimize hydraulic head and solve quadratic programming problems: Amherst, MA, University of Massachusetts Department of Civil and Environmental Engineering, Masters Thesis, 119 p.

Bexfield, L.M., Danskin, W.R., and McAda, D.P., 2004, Simulation-optimization approach to management of ground-water resources in the Albuquerque area, New Mexico, 2006 through 2040: U.S. Geological Survey Scientific Investigations Report 2004-5140, 82 p.

Eggleston, J.R., 2004, Evaluation of strategies for balancing water use and streamflow reductions in the Upper Charles River Basin, Eastern Massachusetts: U.S. Geological Survey Water-Resources Investigations Report 03-4330, 85 p.

Harbaugh, A.W., 2005, MODFLOW-2005, The U.S. Geological Survey modular ground-water model—The Ground-Water Flow Process: U.S. Geological Survey Techniques and Methods 6–A16 [variously paged].

Harbaugh, A.W., Banta, E.R., Hill, M.C., and McDonald, M.G., 2000, MODFLOW-2000, the U.S. Geological Survey modular ground-water model—User guide to modularization concepts and the Ground-Water Flow Process: U.S. Geological Survey Open-File Report 00–92, 121 p.

Hillier, F.S., and Lieberman, G.J., 1990, Introduction to operations research (5th ed.); New York, McGraw-Hill Publishing, 954 p.

Huili, Gong, Menlou, Li, and Xinli, Hu, 2000, Management of groundwater in Zhengzhou City, China: Water Research, v. 34, no. 1, p. 57–62.

Male, J.W., and Mueller, F.A., 1992, Model for prescribing ground-water use permits: Journal of Water Resources Planning and Management, v. 118, no. 5, p. 543–561.

McDonald, M.G., and Harbaugh, A.W., 1988, A modular three-dimensional finite-difference ground-water flow model: U.S. Geological Survey Techniques of Water-Resources Investigations, book 6, chap. A1, 586 p.

McPhee, James, and Yeh, W. W.-G., 2004, Multiobjective optimization for sustainable groundwater management in semiarid regions: Journal of Water Resources Planning and Management, v. 130, no. 6, p. 490–497.

Mehl, S.W., and Hill, M.C, 2005, MODFLOW-2005, The U.S. Geological Survey modular ground-water model—Documentation of shared node local grid refinement (LGR) and the boundary flow and head (BFH) Package: U.S. Geological Survey Techniques and Methods 6–A16, 68 p.

Mehl, S.W., and Hill, M.C., 2007, MODFLOW-2005, The U.S. Geological Survey modular ground-water model—Documentation of the multiple-refined-areas capability of local grid refinement (LGR) and the boundary flow and head (BFH) package: U.S. Geological Survey Techniques and Methods 6–A21, 13 p.

Mueller, F.A., and Male, J.W., 1993, A management model for specification of groundwater withdrawal permits: Water Resources Research, v. 29, no. 5, p. 1359–1368.

Niswonger, R.G., and Prudic, D.E., 2005, Documentation of the Streamflow-Routing (SFR2) Package to include unsaturated flow beneath streams—A modification to SFR1: U.S. Geological Survey Techniques and Methods 6–A13, 48 p.

Prudic, D.E., Konikow, L.F., and Banta, E.R., 2004, A new Streamflow-Routing (SFR1) Package to simulate stream-aquifer interaction with MODFLOW-2000: U.S. Geological Survey Open-File Report 2004–1042, 95 p.

Appendix 1

Data-Input Instructions and Output Files

Contents

Table

Data-Input Instructions

Data-input instructions for GWM-2005 when state variables are not used are fully described in Ahlfeld and others (2009). When state variables are included in a management formulation, the user must prepare a state-variables (**STAVAR**) input file and activate the STA Package in the **GWM** file. Additional changes also may be necessary for the **DECVAR**, **VARCON**, **OBJFNC**, and **SUMCON** files, depending on how the state variables are used in the formulation. All the data-input instructions necessary for the use of state variables are described in this appendix. Keywords used in the data-input files are shown in bold, uppercase and italicized text, such as the keyword *OUT*.

GWM File

The **GWM** file is formatted in a manner similar to the MODFLOW **NAME** file. A series of records are read that have the following format:

Ftype   Fname

Ftype is one of several keywords, and Fname is a path name of the relevant computer file. Except for keyword *OUT* (described below), each of the keywords triggers the reading of a file that will be referred to with the same name as the keyword. The entire record including the Fname entry is limited to 199 characters in length. Comment lines may appear anywhere in the **GWM** file and are indicated by the # character in the first column of the record.

Keywords can be specified in either uppercase or lowercase letters. Keywords may appear in any order except (1) the *OUT* keyword must be the first keyword in the file if it is used, and (2) the *STAVAR* keyword must follow the *DECVAR* keyword if state variables are defined for the problem. The keywords suitable for inclusion in a **GWM** file depend on the type of problem. If the problem is a single model (that is, a simulation without local grid refinement), then only a single **GWM** file is provided. If the problem is multimodel (with local grid refinement), then a **GWM** file is required for the parent model and may be provided for child models. The following keywords are available in GWM-2005:

OUT—a filename for all output from the GWM Process may be assigned here. If the *OUT* keyword is not specified, a default name of "GWM.OUT" will be used, and the output file will be written to the directory in which program execution occurs. The *OUT* keyword is not allowed if the **GWM** file is for a child model.

DECVAR—the Fname associated with this keyword identifies the **DECVAR** file that provides information about the decision variables. For single-model problems, the *DECVAR* keyword is required. For multimodel problems, a **DECVAR** file is provided for every model that includes decision variables. The **GWM** file for at least one model, although not necessarily the parent model, must contain a *DECVAR* keyword.

STAVAR—the Fname associated with this keyword identifies the **STAVAR** file that provides information about the state variables. A **STAVAR** file is optional, but if it is listed, it must follow the **DECVAR** file record. For multimodel problems, a **STAVAR** file is provided for every model that includes state variables.

OBJFNC—the **OBJFNC** file provides information about the objective function. The *OBJFNC* keyword must appear in the **GWM** file for single-model problems and in the parent model **GWM** file for multimodel problems. The *OBJFNC* keyword is not allowed in the **GWM** files of child models of multimodel problems.

VARCON—the **VARCON** file provides information on the lower and upper bounds specified for the decision variables defined in the **DECVAR** file. If the *DECVAR* keyword appears in a **GWM** file, then the *VARCON* keyword must also appear.

SUMCON, **HEDCON,** and *STRMCON*—the **GWM** file may include up to three additional files that provide information about summation constraints, head constraints, and streamflow constraints that are allowed in GWM. None of these keywords are required in a **GWM** file. For multimodel problems, the *SUMCON* keyword can appear only in the parent model, whereas the *HEDCON* and *STRMCON* keywords may appear in the **GWM** files for parent or child models.

SOLN—the **SOLN** file provides information about the solution and output-control parameters. The *SOLN* keyword must appear in the **GWM** file for single-model problems and in the parent model **GWM** file for multimodel problems. The *SOLN* keyword is ignored in the **GWM** files of child models of multimodel problems.

The requirements for the specification of file types in GWM-2005 are summarized in Ahlfeld and others (2009, p. 22–23). Those requirements are unchanged with the addition of the STA Package, except for the use of a **STAVAR** file, as shown in table A1-1.

Table A1-1.   GWM file requirements for simulations with and without local grid refinement (LGR).

File type	Simulation without LGR	Simulation with LGR	
		Parent model	**Child model(s)**
GWM	Required	Required	Optional
DECVAR	Required	Optional[1]	Optional[1]
STAVAR	Optional	Optional	Optional
OBJFNC	Required	Required	None
VARCON	Required	Optional[2]	Optional[2]
SUMCON	Optional	Optional[3]	None
HEDCON	Optional	Optional	Optional
STRMCON	Optional	Optional	Optional
SOLN	Required	Required	None
OUT	Optional	Optional	None

[1]At least one **DECVAR** file and associated **VARCON** file must be specified in either the parent or child models or both.

[2]A **VARCON** file must be specified if a **DECVAR** file is specified for the model.

[3]Constraints specified in a **SUMCON** file in the parent model may reference decision variables defined on the parent grid or any of the child grids.

State Variables (STAVAR) File

This optional file is used to define the state variables of the management problem. State variables represent system state and can be used in the objective function or in linear-summation constraints. Head, streamflow, and change-in-aquifer-storage (storage) variables are the types of state variables currently supported in GWM-2005. Head and streamflow state variables are associated with a specified location and are evaluated at the end of a specified stress period. Head and streamflow state variables must be defined in the **STAVAR** file associated with the parent or child grid within which the state variable is located.

Storage state variables record the change in aquifer storage within a specified region of the model domain over a specified time period. The specified time period is defined by starting and ending stress periods. The time period extends from the beginning of the starting stress period, defined by the SPSTRT input variable, to the end of the ending stress period, defined by the SPEND input variable. The specified region can be any set of model cells selected by the user. It may include the entire model domain or be limited to a portion of the domain, as specified in the CZONE input variable. When the storage state variable is associated with only a portion of the domain, the portion is defined cell-by-cell using the NSVZL, LNUM, and SVZONE input variables. When a multigrid model is used, the **STAVAR** file can be provided for each grid. Storage state variables may be defined over the entirety or just a portion of each individual grid. Storage state variables also can be defined that include change in storage in multiple grids. This is accomplished by assigning the same storage state variable name, SVNAME, in each of the **STAVAR** files that include the state variable. For example, if the storage state variable is intended to cover the entire model domain, a **STAVAR** file would be provided for each grid and would contain an entry specifying the same SVNAME.

The **STAVAR** file consists of five input items:

0.   [#Text]

Item 0 is optional—# must appear in column 1. Item 0 can be repeated multiple times.

1.   IPRN
2.   NHVAR   NRVAR   NSVAR
3.   The following record is read for each of the NHVAR head state variables:
      SVNAME   LAY   ROW   COL   SVSP
4.   The following record is read for each of the NRVAR streamflow state variables:
      SVNAME   SEG   REACH   SVSP
5a.   The following record is read for each of the NSVAR storage state variables:
      SVNAME   SPSTRT   SPEND   CZONE
5b.   The following record is read if CZONE is assigned a value of "ZONE":
      NSVZL
5c-d. The following two records are read NSVZL times:
      LNUM
      SVZONE   (Read using the MODFLOW U2DINT utility subroutine; see Harbaugh, 2005, p. 8–57 through 8–59)

The variables are defined as follows:

Text—is a character variable up to 199 characters long that starts in column 2. Any characters can be included in Text. Lines beginning with # are restricted to the first lines of the file. Text is printed when the file is read.

IPRN—is an integer variable that describes the amount of output written to the **GWM OUT** file. IPRN must be specified as either 0 or 1. When IPRN equals 0, a minimum amount of information about the decision variables is written to the GWM output file; when IPRN equals 1, detailed information about the decision variables is written to the GWM output file.

NHVAR—is an integer variable equal to the number of head-type state variables. NHVAR must be greater than or equal to 0.

NRVAR—is an integer variable equal to the number of streamflow-type state variables. NRVAR must be greater than or equal to 0.

NSVAR—is an integer variable equal to the number of storage-type state variables. NSVAR must be greater than or equal to 0.

SVNAME—is a character variable up to 10 characters long that is a unique name designated for the state variable. For head and streamflow state variables, each name must be unique (that is, the same name cannot be used for more than one variable, or in more than one model). For storage state variables applied to multimodel problems (that is, those using local grid refinement), the same name may appear in more than one **STAVAR** file in order to define a storage state variable that extends over multiple grids. However, within a given **STAVAR** file on one grid of a multimodel problem, the state variable name must be unique. No spaces are allowed in the name. The end of the name is designated by a blank space.

LAY, ROW, and COL—are integer variables equal to the layer, row, and column number of the model cell in which the head-type state variable is located.

SVSP—is an integer variable that indicates the stress period during which the head or streamflow state variable is to be evaluated. To evaluate a head or streamflow state variable for multiple stress periods, define multiple state variables.

SEG and REACH—are integer variables equal to the segment and reach numbers of the model cell in which the streamflow-type state variable is located. The SEG and REACH numbers must correspond to a valid segment and reach as specified in either the STR or SFR input files.

SPSTRT and SPEND—are integer variables equal to the stress periods at which the evaluation of the storage state variable will start and end. The change in storage associated with the state variable will be computed by subtracting the volume of water in a specified portion of the model domain at the beginning of stress period SPSTRT from the volume of water in the same portion at the end of stress period SPEND.

CZONE—is a character variable that describes the portion of the aquifer domain to be included in the storage state variable. Two options are allowed:

ALL—the storage state variable will record the change in water stored in the entire model domain.

ZONE—the storage state variable will record the change in water stored in a portion of the model domain that is defined in subsequent records in the file.

NSVZL—is an integer variable equal to the number of model layers included in the storage state-variable zone. The zone array will be read one layer at a time.

LNUM—is an integer variable equal to the layer number for the storage state-variable zone array.

SVZONE—is a two-dimensional (one layer) zone array that is read using the U2DINT utility subroutine of MODFLOW (see Harbaugh, 2005, p. 8–57 through 8–59). A value will be read for each cell in the model layer. A value of zero indicates the cell is not included in the storage state variable; a value greater than zero indicates the cell will be included in the definition of the storage state variable.

Revised Instructions for Decision Variable (DECVAR) and Decision-Variable Constraints (VARCON) Files

Small modifications have been made to four of the input variables defined in the **DECVAR** and **VARCON** files; these modifications are described below. All other variable definitions in the **DECVAR** and **VARCON** files remain unchanged (see Ahlfeld and others, 2009, p. 23–27).

The following changes were made to variables ETYPE and ESP in the **DECVAR** file to accommodate the expanded definitions of external variables:

ETYPE—is a character variable that indicates the external variable type. ETYPE can be assigned one of six values: IM defines the external variables as a source (import) of water; EX defines the variable as a sink (export) of water; HD defines the variable as a head type; SF defines the variable as a streamflow type; ST defines the variable as a storage type; and GN defines the external variable as a general type. Any combination of external-variable types can be used in a management problem. The value of ETYPE is used by GWM in the output for the external variable. Regardless of the variable type, all external variables are treated as positive-valued variables.

ESP—is a character string up to 120 characters long that indicates the stress periods associated with external variable EVNAME. A single value of the external variable will be determined by GWM for all of the stress periods included in ESP. The total time during which the external variable is active is determined by summing the durations of all the stress periods in ESP. This total time is applied to the objective function only if FNTYP is specified as WSDV (eq. 2a) in the **OBJFNC** file. The string must not contain any blank spaces. Multiple stress periods are indicated by colons (:) or hyphens (-). For example,

   1 indicates that stress period 1 is the only stress period associated with the decision variable,

   1:3 indicates that the flow rate for the external variable is the same for stress periods 1 and 3 (but not 2), and

   1-12 indicates that the flow rate for the external variable is the same for stress periods 1 through 12.

In addition, the following changes were made to variables EVMIN and EVMAX in the **VARCON** file, also to accommodate the expanded definitions of external variables:

EVMIN and EVMAX—are real variables equal to the minimum (EVMIN) and maximum (EVMAX) values allowed for the external decision variable. Because external variables are defined as positive-valued variables, values greater than or equal to 0 must be specified for EVMIN and EVMAX. EVMIN must be less than or equal to EVMAX. Note that a nonzero value of EVMIN implies that the decision variable has been associated with a binary variable in the **DECVAR** file. If the decision variable is not associated with a binary variable, the nonzero value of EVMIN is ignored by GWM and EVMIN is set to zero. The user can specify a nonzero lower bound for an external decision variable not associated with a binary variable by use of a summation constraint (see description of **SUMCON** file).

Revised Instructions for Objective Function (OBJFNC) File

The **OBJFNC** file is used to define the objective function that is to be maximized or minimized and the coefficients for each decision or state variable in the objective function. Input for this file was slightly modified for the addition of the STA Package to GWM-2005. Variables that were added are NSVOBJ in item 3 and SVNAME and SVOBJC in item 7; the definition of variable FNTYP also has been expanded. All other variable definitions remain unchanged (see Ahlfeld and others, 2009, p. 25–26).

The **OBJFNC** file includes seven input items:
0.   [#Text]
Item 0 is optional—# must appear in column 1. Item 0 can be repeated multiple times.
1.   IPRN
2.   OBJTYPE   FNTYP
3.   NFVOBJ   NEVOBJ   NBVOBJ   [NSVOBJ]
4.   The following record is repeated for each of the NFVOBJ flow-rate decision variables:
      FVNAME   FVOBJC
5.   The following record is repeated for each of the NEVOBJ external decision variables:
      EVNAME   EVOBJC
6.   The following record is repeated for each of the NBVOBJ binary decision variables:
      BVNAME   BVOBJC
7.   The following record is repeated for each of the NSVOBJ state variables:
      SVNAME   SVOBJC

The new variables are defined as follows:

FNTYP—is a character variable used to define the type of objective function. Three options are allowed:
   WSDV—the objective function takes the form of a weighted sum of decision variables (eq. 2a). Weighting is automatically applied by multiplying flow-rate, external, and state variables by the length of time the variable is active. The length of time is determined by summing the length of all stress periods during which the decision variable is active, as defined in the associated definition of the variable. This form of the objective function is commonly used to convert variables that have units of flow rate to variables with units of volume. Additional weighting is applied to the variables by the user-specified values of the cost/benefit coefficients (FVOBJC, EVOBJC, BVOBJC, and SVOBJC).

   USDV—the objective function takes the form of an unweighted sum of decision variables (eq. 2b). The variables are not weighted by time. Additional weighting is applied to the variables by the user-specified values of the cost/benefit coefficients (FVOBJC, EVOBJC, BVOBJC, and SVOBJC).

MSDV—the objective function takes the form of a sum of decision variables with mixed weighting (eq. 2c). Flow-rate decision variables are weighted by the duration of their activity, but external and state variables are not weighted by their duration of activity. Additional weighting is applied to the variables by the user-specified values of the cost/benefit coefficients (FVOBJC, EVOBJC, BVOBJC, and SVOBJC).

NSVOBJ—is an optional integer variable equal to the number of state variables in the objective function. NSVOBJ must have a value that is less than or equal to the sum of NHVAR, NRVAR, and NSVAR specified in the **STAVAR** file.

SVNAME—is a character variable up to 10 characters long that is one of the state-variable names defined in the **STAVAR** file. A state-variable name can be listed only once in the **OBJFNC** file.

SVOBJC—is a real variable that is a coefficient associated with each state variable SVNAME.

Revised Instructions for Linear-Summation Constraints (SUMCON) File

The **SUMCON** file is used to define linear relations among decision variables. The only change that was made to this file for the addition of the STA Package is a slight modification to the definition of variable GVNAME, which is shown below. All other variable definitions remain unchanged (see Ahlfeld and others, 2009, p. 27–28).

GVNAME—is a character variable up to 10 characters long that is one of the decision-variable names defined in a **DECVAR** file or one of the state-variable names defined in a **STAVAR** file. Any combination of flow-rate (FVNAME), external (EVNAME), binary (BVNAME), or state (SVNAME) variables may be defined in a constraint. The user must ensure that the variables included are logically consistent.

State Variables in the Output File

State-variable input information is echoed to the **OUT** file produced by GWM. Once a state variable is assigned a name, it is referred to by that name in all references to it in the objective function and constraints. The output for the optimal solution printed to the **OUT** file will include the optimal values for the state variables, which are listed following the optimal values for the decision variables. Just as with decision variables, the contribution of the state variable to the objective function is reported along with totals for state-variable values and contributions to the objective. Because state variables are computed using a Taylor-series linearization within the optimization solution algorithm, the values of the state variables are also reported in the final flow-process simulation portion of the output. These values have been directly computed by the flow process using the optimal values of the flow-rate decision variables. The user should compare these values with those reported as part of the optimal solution. Substantial differences may indicate the presence of nonlinear responses or numerical precision issues that should be corrected.

Range analysis is described in detail in Ahlfeld and others (2005). In brief, it describes the range over which the values of the right-hand side of each constraint and of the objective-function coefficients may vary without changing the basis. The range analysis also reports the variable or constraint slack, which will enter and leave the basis if the basis does change. The range-analysis portion of the output is unchanged from prior versions of GWM. As a result, state variables are not explicitly expressed in the range analysis. Only the fundamental variables (the flow-rate and external decision variables) are considered in the range analysis. The relation of these variables to the state-variable values must be inferred from the state-variable response coefficients.

References Cited

Ahlfeld, D.P., Baker, K.M., and Barlow, P.M., 2009, GWM-2005—A Groundwater-Management Process for MODFLOW-2005 with Local Grid Refinement (LGR) capability: U.S. Geological Survey Techniques and Methods 6-A33, 65 p.

Ahlfeld, D.P., Barlow, P.M., and Mulligan, A.E., 2005, GWM—A Ground-Water Management Process for the U.S. Geological Survey modular ground-water model (MODFLOW-2000): U.S. Geological Survey Open-File Report 2005–1072, 124 p.

Harbaugh, A.W., 2005, MODFLOW-2005, The U.S. Geological Survey modular ground-water model—The Ground-Water Flow Process: U.S. Geological Survey Techniques and Methods 6–A16 [variously paged].

Appendix 2

Programmers' Guide to Implementation of State Variables in GWM-2005

The STA Package consists of five procedures: STA AR for reading state-variable information from the **STAVAR** input file and echoing to the **OUT** file; STA OS, which assembles state-variable values from computed GWF Process information and assigns values to the state-variable array; STA FP, which computes the state-variable response matrix by differencing state-variable array values; STA FPR, which allows the state-variable response matrix to be read from a file; and STA OT, which writes state-variable information to the output file.

The STA Package procedures are called from the main program, the GWM Basic Package (BAS), the Objective Function Package (OBJ), the Response Matrix Solution Package (RMS), and the Summation Constraints Package (SMC).

The STA OS procedure is called from the main program within the time-step loop of the GWF Process to provide STA the information needed to assemble the state-variable array. In the Basic Package, procedure BAS AR calls the STA AR procedure if a state-variable file (**STAVAR**) is indicated in the GWM file. In the OBJ Package, procedure OBJ AR reads and processes state-variable objective-function information, procedure OBJ FM uses the state-variable response matrix to replace any state variables in the objective function with the corresponding linearization in terms of flow-rate decision variables, and procedure STA OT writes the optimal state-variable results to the output file. In the RMS Package, procedure RMS PL calls STA FPR to read or print the response matrix to/from a file, procedure RMS FP calls STA FP to calculate the state-variable response coefficients, procedure RMS FM writes the state-variable response matrix to a file, and procedure RMS OT calls STA OT to write information to the output file. In the SMC Package, procedure SMC AR uses state-variable name information, procedure SMC FM uses the state-variable response matrix to replace any state variables in the summation constraints with the corresponding linearization in terms of flow-rate decision variables, and procedure SMC OT uses state-variable information for output.